Concrete Folded Plate Roofs

Concrete Folded Plate Roofs

C B Wilby

PhD, BSc, CEng, FICE, FIStructE
Consultant and Emeritus Professor
The University of Bradford, UK

CRC Press
Taylor & Francis Group
Boca Raton London New York

CRC Press is an imprint of the
Taylor & Francis Group, an **informa** business

CRC Press
Taylor & Francis Group
6000 Broken Sound Parkway NW, Suite 300
Boca Raton, FL 33487-2742

First issued in paperback 2019

Copyright © 1998 C B Wilby.
CRC Press is an imprint of Taylor & Francis Group, an Informa business

The right of C. B. Wilby to be identified as the author of this work
has been asserted in accordance with the Copyright, Designs and
Patents Act 1988

No claim to original U.S. Government works

ISBN-13: 978-0-415-50318-1 (hbk)
ISBN-13: 978-0-367-86686-0 (pbk)

Visit the Taylor & Francis Web site at
http://www.taylorandfrancis.com

and the CRC Press Web site at
http://www.crcpress.com

British Library Cataloguing in Publication Data
A catalogue record for this book is available from the British Library

Library of Congress Cataloguing in Publication Data
A catalogue record for this book is available from the Library of Congress

To Jean, Charles (Anthony), Chris, Mark and Dr I. Khwaja

Contents

List of plates

Plate 1   Folded plate roofs for a teacher training college, Leeds, Yorkshire, UK. Each of the vaults shown is 3.35 m (11 ft) wide by 19.84 m (65 ft) span, whilst similar vaults over the gymnasium nearby are of 21.85 m (71 ft 8 in). The plates are 102 mm (4 in) thick. The columns are held together at their heads, at the gables, by 229 mm (9 in) wide by 152 mm (6 in) deep post-tensioned prestressed concrete ties. These were designed by the author.

Plate 2   Turin Exhibition Hall, Italy, designed by P.L. Nervi.

Plate 3   Sears Store, Tampa, Florida, USA. (Courtesy of Portland Cement Association, USA.)

Plate 4   Air terminal, Minneapolis, USA. (Courtesy of Portland Cement Association, USA.)

Plate 5   School in Albuquerque, USA. (Courtesy of Portland Cement Association, USA.)

Plate 6   Swimming pool, Nottingham, UK. (Courtesy of Charles A. Wilby.)

Plate 7   Entrance to Bradford College, Yorkshire, UK.

Plate 8   High School, Scottsdale, Arizona, USA. (Courtesy of Portland Cement Association, USA.)

Plate 9   Bank in Colorado, USA. (Courtesy of Portland Cement Association, USA.)

Plate 10   Chemical works, ICI Hyde, Cheshire, UK. These are reinforced concrete barrel vault roofs but show precast concrete lights comprising glass lenses framed in reinforced concrete which can be similarly used for folded plates. The wall lighting was similar so that no painting maintenance was envisaged as this was difficult to organise in the works' very short holiday.

Plate 11   These are the folded plates to the gymnasium mentioned in the caption to Plate 1. They show roof lighting similar to that shown in Plate 10 and similar end gable wall lighting.

Plate 12   Restaurant, Novato, California, USA. (Courtesy of Mark Stainburn Wilby.)

Plate 13   St Paul's Church, Lorrimore Square, Walworth, London, UK. (Courtesy of British Cement Association, Crowthorne, Berkshire, UK.)

Plate 14   Nuestra Señora del Rosario High School, Spain. (Courtesy of British Cement Association, Crowthorne, Berkshire, UK.)

Plate 15   IBM Pavilion, Brussels Exhibition, Belgium. (Courtesy of British Cement Association, Crowthorne, Berkshire, UK.)

Plate 16   American Concrete Institute Office, Detroit, Michigan, USA. The folded plates were 114 mm (4.5 in) thick. (Courtesy of British Cement Association, Crowthorne, Berkshire, UK.)

Plate 17   Sonoma Development Centre, Eldridge, California, USA. (Courtesy of Mark Stainburn Wilby.)

Plate 18   Coventry Pedestrian Precinct, UK.

Preface

Folded plates are sometimes called hipped plates and, in Germany, *Faltwerke*. The principle was first used in Germany by Ehlers, in 1924, not for roofs but for large coal bunkers, and Ehlers published a paper on their structural analysis in 1930. Then, in 1932 Gruber published an analysis in German. In the next few years many Europeans – Craemer, Ohlig, Girkmann and Vlasov (1939) amongst them – made contributions to this subject. However, the European theories were generally too complex and arduous for design use.

Since about 1945, simplified methods have been developed in the USA by Winter and Pei (1947), Gaafar (1953), Simpson (1958), by Whitney (1959) adapting the method by Girkmann, and by Parme (1960). In 1974, Professor Haas told the author that, in mainland Europe, 'we go by Girkmann (1948)'. In addition, Yitzhaki (1959) of Israel had a book published in English depicting his analysis of folded plates.

The ASCE Task Committee on Folded Plate Construction, 1963, recommended a modified version of Gaafar's method for design office use. Before 1963, the author was using Gaafar's method for designing folded plates. Then, from 1963, he taught MSc-course students the methods of Gaafar, then Simpson, and then, when it became easy to solve simultaneous equations on a new in-house desk top computer, Parme. In addition, some of the students wrote computer programs for Parme's method. With Parme's kind permission and his assurance that his method was correct, as it had been checked by computer, the author reproduced his method – as it gives directly simultaneous equations that can quickly be solved by computer – in two of his general books for students of concrete structures and materials, published by Newnes-Butterworths then Cambridge University Press.

The present book considers schemes of folded plates of practical use to designers for covering large or small areas economically and gives design tables that can be used easily and quickly without the designers actually needing to know how to analyse folded plates. Examples are given on the use of the tables. These examples are produced for those using USA (British Imperial) as well as SI units. The text gives the designer practical information on construction, materials, insulation, weather proofing, etc., useful for these types of roofs.

The book is similar to the author's book *Concrete Shell Roofs* (co-authored with Dr I. Khwaja, Elsevier, 1977), which gives design tables for cylindrical shell roofs, and his book *Reinforced Concrete Conoidal Shell Roofs* (co-authored with Dr M.M. Naqvi, Chapman & Hall, 1973), which gives design tables for conoidal shell roofs, and his book *Concrete Dome Roofs* (Longman, 1993), which gives design tables for domes. The book *Design Graphs for Concrete Shell Roofs*, also by the author (Elsevier, 1980), similarly helps the designer, in this case using graphs instead of tables.

There are different opinions on what should be taught in the way of design and design office work in undergraduate and graduate courses. Some civil engineers after graduation only spend a few months to one year on structural concrete design in practice and are therefore restricted to simple designs, beams, slabs and columns, and it may be that their employer does not have any more complex designs at that time. For these reasons designers with many years experience may never design folded plates or shells, particularly as some offices would only entrust perhaps a certain individual to specialise in such designs. Therefore, the author considers that it is desirable for students to learn some of their basic reinforced concrete detailing by having the opportunity they may never get in practice, of designing folded plate structures, domes or other shell roofs using the design tables or graphs mentioned in the author's previous books.

Traditionally, one is supposed to tackle difficult topics in a degree course. Many years ago an eminent Professor said that one should not teach anything on a degree course that could subsequently easily be acquired. This is a rather severe statement but has a recognisable ideology. However, the principal market for this book will be practical designers and their libraries and the libraries of universities and technological institutions around the world.

Folded plate roofs enable roofs of durability, large spans for reinforced concrete, clean lines, and they are aesthetically pleasing to many architects.

There does not seem to be a book particularly useful to modern practical designers devoted entirely to folded plate roofs and giving design tables that can be used without the designer having to spend a long time endeavouring to study the necessary complex basic theories/analyses. This book provides this service and it indicates that any code of practice might be used internationally. This is done by giving examples using British and USA (ACI) Codes. Although the book uses SI units throughout, it provides adequately for those using British Imperial/USA units.

The author has had a similar experience here to when he wrote his book on domes. The examples are long and those published that can be followed through in complete detail are few. In addition, when one studies some of them in detail one finds mistakes. The author thought it would be just a matter of programming known technology and producing tables which, from his considerable practical experience, he considered to be useful to practical designers. In fact, the author considers that the only example in the past literature which is definitely correct is the one by H. Simpson, because the same example was solved by G.G. Goble using his method and was also solved by E. Traum using yet another method, the agreement in the results being very good. In addition, the author has supervised many part-time MSc students designing folded plates of differing dimensions using Simpson's method.

In writing this book, the author used a program kindly lent by a colleague, then used one he wrote himself for Parme's method but could not get the results to agree with Simpson's. He spent an enormous amount of time trying to solve this discrepancy and eventually sent his observations to the Portland Cement Association of the USA, but they had inadequate staff to allocate to this problem. The author then spent another immense amount of time trying to find the trouble. He feels that Parme is very capable and his theoretical work is probably absolutely correct, but it does result in numerous very long equations and, in being reproduced in the *PSA Bulletins*, it only needs a small error somewhere to upset everything. The author considers he has found some of these errors.

In an attempt to find a different method, the author approached Dr I. Khwaja – a former colleague at Bradford University, who is now in London where he has been working in Building Control for many years – who once wrote a program based on Gibson's method for some folded plate design tables, which the author published in *Concrete for Structural Engineers* (Newnes-

Butterworths, 1977) and which checked with Simpson's example. Dr I. Khwaja very kindly wrote another program directly dealing with plates and, eventually, this checked accurately with examples by Simpson, Yitzhaki (his book), Thadani (*Indian Concrete Journal*, 1957) and J. Born (his book, Crosby Lockwood) and was used for the tables of this book. The author is greatly indebted to Dr Khwaja for his kind help.

Professor C.B. Wilby
Harrogate and Scarborough, North Yorkshire, UK

Disclaimer of warranty and conditions of sale

Acknowledgements

The author is tremendously indebted to Dr I. Khwaja, previously of Bradford University and then Building Control in London, for his programming help, with the production of the design tables and for his generous help and discussions of complex problems over many years.

The author is also indebted for the most useful help, kindly and pleasantly given, by Anne Costigan of the Bradford University Library, and to the Library itself.

Nomenclature

B, F0, F1, F2, F3, M0, M1, M2, M3: see Section 3.7. Note that 0 refers to the outside face and not the top of the beam for all propped beam cases.

C = compression force

d = depth of edge beam

E_c = Young's modulus for concrete

h = sloping width of plate

I = second moment of area of section

l = length of simply supported span

L = span of folded plate

M = bending moment

t = thickness of plate

T = tension force

w = load per unit length

X = unknown distance

α_e, α_f, x_1, z_1, K are explained in Section 4.2.2

θ = angle of principal stress

1
Practicalities

1.1 Uses

Folded plates have been used on various buildings, for instance storage buildings, swimming pools, gymnasia, offices, centres, entrances to buildings and tunnels – for examples see Plates 1–18. Sometimes industrialists like to have the facility to hang unpredicted miscellaneous light loads from anywhere under a roof and regard the structural steelwork as inherently providing this facility. Because of this requirement the author designed the shells shown in Plate 10 to have a network of numerous cadmium-plated steel bolts placed through holes in the shells and through steel anchorage plates of 152 mm (6 in) square on the top surface of the shells. Each bolt protruded out of the soffit of the shell so that just about anything could be screwed on to it at some future date. The nuts and plates were covered with a 50 mm (2 in) layer of vermiculite insulation on the top of the shell, waterproofed with three layers of built-up roofing felt. This facility can similarly be applied to the plates of folded plate roofs.

1.2 Advantages and disadvantages

Because they are of concrete, such roofs have inherent resistance to fire, deterioration and to atmospheric corrosion. They allow large spans to be achieved in structural concrete. This allows flexibility of planning and mobility beneath. Where ground conditions require expensive piled foundations the reduced number of supporting columns can be an economic advantage. For large spans in structural concrete folded plates compete with barrel vault roofs. The plates are required to be thicker than the shells, and there are more firms who will tackle constructing them without excessive prices, increasing competition and sometimes making the cost more competitive than for cylindrical shells.

In the UK there are firms specialising in shuttering (formwork), which are very skilled at curved shuttering, and the author has experienced a contract where letting out the curved shuttering to a specialist was no more expensive than the price we were charging for the nearby flat shuttering at a lesser height. However, in the author's experience in the UK, there are many contractors inexperienced at curved work who will happily quote for folded plates. For example, many smaller firms with less overheads will compete for folded plates but not shell roofs.

Some architects prefer the aesthetics of folded plates to curved shell roofs. Folded plates provide good quality robust roofs. They are, however, usually more expensive than roofs much lighter in weight comprising roof sheeting (even heavier wood wool slabs with a thin

sand/cement screed and three layers of built-up roofing felt, the top layer being mineral finished)
supported by purlins and frames of structural concrete.

1.3   Practical types

Plates 1–18 show various folded plates. Figure 1.1 shows a useful system of folded plates, and
Fig. 1.2 shows the cross-section of a type of folded plate roof commonly analysed in USA publications (e.g. Refs 1.1 and 1.2). Figure 1.3 shows a 'trough' type which, for long spans, accom-
modates the reinforcement in the valleys more easily. Figures 1.4 and 1.5 show two North-light
types. The slope of the glazing of a North-light depends upon one's latitude, theoretically tending
from 90° at the equator to 0° at the poles.

Figure 1.6 shows a similar profile to Fig. 1.1. For longer spans, sometimes the profiles of Figs
1.7, 1.8 and 1.9 are preferred to the profiles of Figs 1.5, 1.6 and 1.1 respectively.

In the cases of Figs 1.1, 1.3, 1.5, 1.6, 1.7, 1.8 and 1.9, the end valleys usefully provide gutters
for rainfall. If there is a wall beneath an end valley a suitable joint will be necessary to allow the
valley to deflect in accordance with its design and so as not to damage the wall below, which
might be of brickwork, concrete blockwork, glazing, glass screening, timber, etc. This joint needs
to be weathertight against horizontal wind-driven rain in, for example, the UK, and no doubt
against wind-driven sand or dust in certain hotter climates. This deflection problem can be dealt
with alternatively by propping the end valley with a row of columns as shown in Fig. 1.10. This
also reduces the reinforcement required at this fold. The columns are best when beneath vertical
edge beams such as shown in Figs 1.2 and 1.5 to 1.10 for structural and reinforcement detailing
considerations.

1.4   Design and analysis

Folded plates are sometimes called hipped plates and, in Germany, *Faltwerke*. The principle was
first used in Germany by Ehlers, in 1924, not for roofs but for large coal bunkers and he published
a paper on the structural analysis in 1930. Then, in 1932 Gruber published an analysis in German.

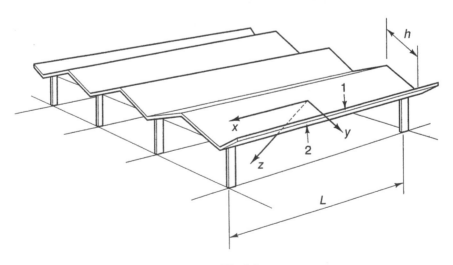

Fig. 1.1

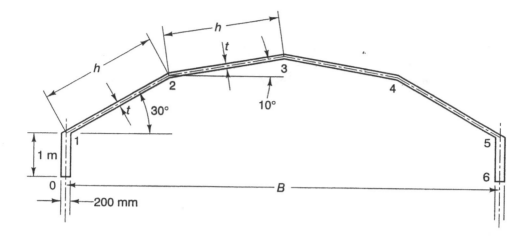

Fig. 1.2

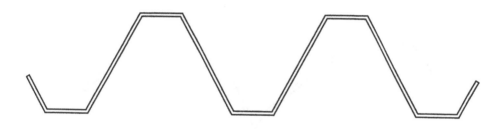

Fig. 1.3

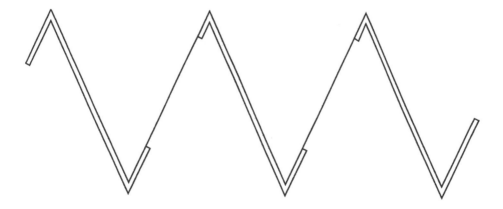

Fig. 1.4

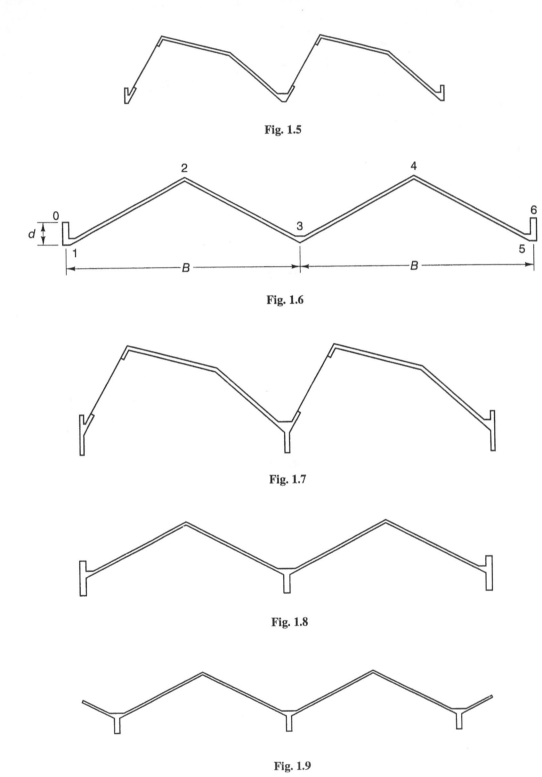

Fig. 1.5

Fig. 1.6

Fig. 1.7

Fig. 1.8

Fig. 1.9

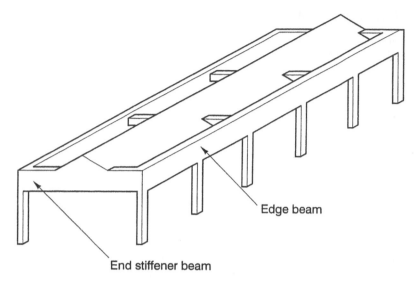

Edge beam

End stiffener beam

Fig. 1.10

In the next few years many Europeans – Craemer, Ohlig, Girkmann and Vlasov (1939) amongst them – made contributions to this subject. The European theories were generally complex and arduous for designer use. Since about 1945 simplified methods have been developed in the USA by Winter and Pei (1947), Gaafar (1953), Ref. 1.3, Simpson (1958), Ref. 1.4, by Whitney (1959) adapting the method by Girkmann, by Traum (1959), Ref. 1.5, by Parme (1960), Ref. 1.6, and by Goble (1964), Ref. 1.7. In 1974, Professor Haas, of the Netherlands, told the author that in mainland Europe 'we go by Girkmann (1948)'. Also Yitzhaki (1959) of Israel had a book published in English depicting his analysis of folded plates. The ASCE Task Committee on Folded Plate Construction, 1963, recommended a modified version of Gaafar's method for design office use. Before 1963, the author used Gaafar's method for designing folded plates. Then, from 1963, he taught MSc students the methods of Gaafar, then Simpson, and then, when it was easy to solve simultaneous equations on a new in-house desktop computer, Parme (Wilby 1983), Ref. 1.8. As a consequence, some of the students wrote computer programs for Parme's method. The author has since written to the PSA throwing doubt on Parme's method as he cannot get it to agree with the example by Simpson, with which Traum and Goble agree.

Methods suitable for computers can be found in works by Goldberg *et al.* (1957, 1964), Refs 1.9 and 1.10, Thadani (1957) and Tamhankar and Jain (1965), Refs 1.11 and 1.12, Gibson and Gardner (1965), Ref. 1.13, and Lo and Scordelis (1969), Ref. 1.14.

The present book considers schemes of folded plates of practical use to designers for covering large or small areas economically and gives design tables that can be used easily and quickly without the designers needing to know how to analyse folded plates. Examples are given on the use of the tables for designers to follow. These examples are produced for those using USA (British Imperial) as well as SI units. The book is similar to the author's book Ref. 1.15 which gives design tables for domes, his book Ref. 1.16, which gives design tables for cylindrical shell roofs and his book Ref. 1.17, which gives design tables for conoidal shell roofs. The book by the author Ref. 1.18 helps the designer similarly but with graphs instead of tables.

1.5   Units of measurement

The units used in this book are SI and USA (British Imperial). To help readers internationally conversions are given in Table 1.1.

Table 1.1

British Imperial	USA	SI	Metric
1 ton	1 long ton	9.964 kN	1016.0 kg
2000 lb	1 short ton	8.896 kN	907.1 kg
0.9843 ton	2205 lb	9.807 kN	1 tonne
1 lb	1 lb	4.448 N	0.4536 kg
1000 lb	1 kip	4.448 kN	453.6 kg
1 in	1 in	25.4 mm	2.54 cm
1 ft	1 ft	0.3048 m	30.48 cm
1 lb/ft	1 lb/ft	14.58 N/m	1.488 kg/m
1 lb/in^2	1 psi	6.895 kN/m^2	0.07031 kg/cm^2
1 lb/in^2	1 psi	6.895 kPa	0.07031 kg/cm^2
1000 lb/in^2	1000 psi	6.895 N/mm^2	70.31 kg/cm^2
1000 lb/in^2	1 ksi	6.895 MPa	70.31 kg/cm^2
1 lb/ft^2	1 lb/ft^2	0.04788 kN/m^2	4.882 kg/m^2
1 lb/ft^2	1 lb/ft^2	47.88 Pa	4.882 kg/m^2
1 ton/ft^2	2240 psf	107.3 kN/m^2	10940 kg/m^2
1 ton/ft^2	2240 psf	107.3 kPa	10940 kg/m^2
1 lb/ft^3	1 lb/ft^3	0.1571 kN/m^3	16.02 kg/m^3

Notes:
1. psf = lb/sq.ft.
2. Pa = Pascal = N/m^2.
3. Use in some European countries: 1 tonne = 1000 kg.
4. The terms 'force' and 'mass' have not been used above, and acceleration due to gravity = 9.807 m/s^2.
5. German literature for a short period before the use of SI units: 1 kilopond = 1 kg force and 1 Mp = 1 megapond = 1 metric tonne force = 1000 kg force (this information courtesy of Professor H. Reiffenstuhl of Austria).

References

1.1   Billington, D.P. (1965) *Thin Concrete Shell Structures*, McGraw-Hill, New York, USA.
1.2   Ramaswamy, G.S. (1984) *Design and Construction of Concrete Shell Roofs*, Krieger, Malabar, Florida, USA.
1.3   Gaafar, I. (1953) Hipped plate analysis considering joint displacements, *Proc Am Soc Civil Engrs*, April.
1.4   Simpson, H. (1958) Design of folded plate roofs, *Proc Am Soc Civil Engrs*, January.
1.5   Traum, E. (1959) The design of folded plates, *Proc Am Soc Civil Engrs, J Structural Div*, October.
1.6   Parme, A.L.L. (1960) Direct solution of folded plate concrete roofs, *Advanced Engineering Bulletin 3*, Portland Cement Association, USA.
1.7   Goble, G.G. (1964) Analysis of folded plate structures, *Proc Am Soc Civil Engrs, J Structural Div*, February.
1.8   Wilby, C.B. (1983) *Concrete Materials and Structures*, Cambridge University Press, Cambridge, UK, and New York, USA.

1.9   Goldberg, J.E. and Leve, H.L. (1957) Theory of prismatic plate structures, *Int Ass Bridge and Structural Engg.*

1.10 Goldberg, J.E., Glauz, W.D. and Setlur, A.V. (1964) Computer analysis of folded plate structures, *Int Ass Bridge and Structural Engg*, Rio de Janeiro.

1.11 Thadani, B.N. (1957) The analysis of hipped plate structures by influence coefficients, *Indian Concrete J*, April.

1.12 Tamhankar, M.G. and Jain, R.D. (1965) Computer analysis of folded plates, *Indian Concrete J*, October.

1.13 Gibson, J.E. and Gardner, N.J. (1965) Investigation of multi-folded plate structures, *Proc Inst Civil Engrs*, May.

1.14 Lo, K.S. and Scordelis, A.C. (1969) Finite segment analysis of folded plates, *Proc Am Soc Civil Engrs, J Structural Div*, May.

1.15 Wilby, C.B. (1993) *Concrete Dome Roofs*, Longman Scientific and Technical, London, and John Wiley, New York.

1.16 Wilby, C.B. and Khwaja, I. (1977) *Concrete Shell Roofs*, Applied Science Publishers, London.

1.17 Wilby, C.B. and Naqvi, M.M (1973) *Reinforced Concrete Conoidal Shell Roofs*, Cement & Concrete Association, London.

1.18 Wilby, C.B. (1980) *Design Graphs for Concrete Shell Roofs*, Applied Science Publishers, London.

2

Analysis used for the design tables

The author published some very limited design tables in Ref. 2.1. These were produced by his colleague Dr I. Khwaja writing computer programs to solve the analytical method suggested by Gibson, Ref. 2.2 as his 'degenerate theory'. This theory uses cylindrical shell theory and flattens out these shells (degenerates them) and tilts them to make systems of folded plates. For the cylindrical shell theory, Dr Khwaja used the very accurate, comprehensive and complex DJK elastic analysis, Ref. 2.3.

Then the author programmed Parme's method, Ref. 2.4, but could not obtain results which agreed with the example given by Simpson, Ref. 2.5. Simpson's example must be correct as it was also solved by different methods by Traum, Ref. 2.6, and Goble, Ref. 2.7. The program written by Khwaja, however, based on Gibson's method, agreed with Simpson's example.

Scordelis, Ref. 2.8 has examined various theories for folded plate structures, including the classical plate theory, and has confirmed their validity by confirming the experimental and theoretical results obtained from the investigation of a multi-folded-plate structure made from light alloy sheet. Gibson, Ref. 2.9 compared the theoretical results using his 'degenerate theory' with those of Scordelis for the same structure and found that agreement was excellent. Further agreement between this 'degenerate theory' and experimental values from a test on a different type of folded plate structure was found by Gibson (1968) and he concluded that stress analysis using this 'degenerate theory' yields results equal to those of the classical elastic theory.

After an immense amount of work finding some mistakes in Parme's publication, the results still did not agree with Simpson's example – and unfortunately the author had intended using Parme's method to produce the tables for this book. The author approached Dr Khwaja to see if he still had his program based on Gibson's method. After several complications, Dr Khwaja was kind enough to write a program he devised for plates, not 'Gibson degenerated shells', and this eventually checked Simpson's example (which agrees with the methods of Traum, Ref. 2.6 and Goble, Ref. 2.7) and examples given in books by Yitzhaki of Israel and J. Born and by Thanani (Ref. 2.10), the ASCE Committee Report (Ref. 2.11), Meek (Ref. 2.12) and Gaafar (Ref. 2.13). The tables have kindly been produced by Dr Khwaja who has worked, on and off, on shell structures with the author for about three decades, and in whom the author has great confidence. He is the author's co-author of Ref. 2.14 and was tremendously helpful in producing Ref. 2.15.

The author has worked with shell structures continuously, amongst other activities concerning plain, reinforced and prestressed concrete, since 1952.

References

2.1   Wilby, C.B. (1977) *Concrete for Structural Engineers*, Newnes-Butterworths, London.

2.2   Gibson, J.E. and Gardner, N.J. (1965) Investigation of multi-folded plate structures, *Proc Inst Civil Engrs*, May.

2.3   Jenkins, R.S. (1947) Theory and design of cylindrical shell structures, O.N. Arup Group of Consulting Engineers, London.

2.4   Wilby, C.B. (1983) *Concrete Materials and Structures*, Cambridge University Press, Cambridge, UK, and New York, USA.

2.5   Simpson, H. (1958) Design of folded plate roofs, *Proc Am Soc Civil Engrs*, January.

2.6   Traum, E. (1959) The design of folded plates, *Proc Am Soc Civil Engrs, J Structural Div*, October.

2.7   Goble, G.G. (1964) Analysis of folded plate structures, *Proc Am Soc Civil Engrs, J Structural Div*, February.

2.8   Scordelis, A.C. (1961) Experimental and analytical study of folded plate structures, *Proc Am Soc Civil Engrs*, December.

2.9   Gibson, J.E. (1968) *The Design of Shell Roofs*, Spon, London.

2.10 Thadani, B.N. (1957) The analysis of hipped plate structures by influence coefficients, *Indian Concrete J*, April.

2.11 ASCE Committee Report (1964) *J Structural Division*, February.

2.12 Meek (1963) ASCE, *J Structural Division*, June.

2.13 Gaafar, I. (1953) Hipped plate analysis considering joint displacements, *Proc Am Soc Civil Engrs*, April.

2.14 Wilby, C.B. and Khwaja, I. (1977) *Concrete Shell Roofs*, Applied Science Publishers, London.

2.15 Wilby, C.B. (1980) *Design Graphs for Concrete Shell Roofs*, Applied Science Publishers, London.

3

Factors used in the design tables

3.1 Types of folded plates used in the design tables

Folded plates of the cross-section shown in Fig. 1.6 are very useful as they are simple to construct, can span considerable distances and cover large or small areas. Their upstanding edge beams and internal valleys provide gutters for rainfall drainage and, if they are to be propped with columns, as in Fig. 1.10, the vertical edge beams are good for local structural, deflection and reinforcement reasons.

In the UK it is not common to provide roofing beyond the walls of a building, as can be the case for folded plates with a cross-section as shown in Fig. 1.1; presumably it is considered to be wasteful of the restricted built-up areas allowed by planning authorities. In France, for example, it is quite common to shield external walls by roof overhangs; seemingly to give shade to windows and, by shielding the external walls from rain, to allow them to be less weather resistant and therefore inexpensive. For example, walls of about 225 mm (9 in) width would often comprise burnt clay hollow blocks set in mortar with no continuous cavity, whereas in the UK the wall would commonly be of 280 mm (11 in) brickwork with a continuous 50 mm (2 in) cavity.

Although folded plates mainly seem to have natural lighting via the walls of the building and sometimes their gables, supplemented by suspended artificial lighting, glass lenses can be cast into the plates if necessary, often using precast panels, with reinforced concrete between the glass lenses. The reinforcement can be designed to give at least the strength required at their locality. When lenses are cast into the folded plates, which are usually at least 100 mm (4 in) thick, the glass lenses can be undesirably thick when they are the thickness of the plate.

Alternatively, circular or rectangular openings can be provided bounded with kerbs strong enough to make up for the strength loss locally due to the opening, to carry glazing (Fig. 3.1), transparent plastic domes (Fig. 3.2), or precast slabs (panels) of reinforced concrete containing lenses (Fig. 3.3).

Many shells were designed in the UK where openings for this type of lighting, up to 1.219 m (4 ft) maximum dimension, were used without altering the structural design of the shells, although at that time it would not have been possible to cater for the holes in the roof in the analysis. In addition, the openings in a shell were not to exceed 9% of the plan area. They must be placed so that clear transverse bands of concrete are left between them, and shells up to 9.144 m (30 ft) width were restricted to two, and shells over that width to four, longitudinal rows of lights. Figures 3.4 and 3.5 show examples of the former and latter respectively for folded plates. The author suggests that this experience, which was found satisfactory for cylindrical shells, should be acceptable for folded plates.

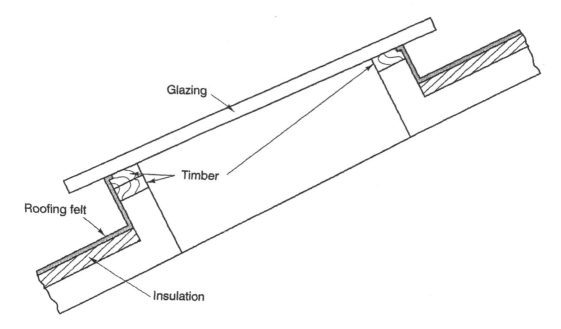

Glazing

Timber

Roofing felt

Insulation

Fig. 3.1

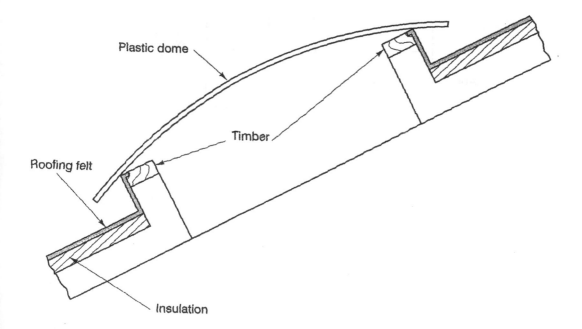

Plastic dome

Timber

Roofing felt

Insulation

Fig. 3.2

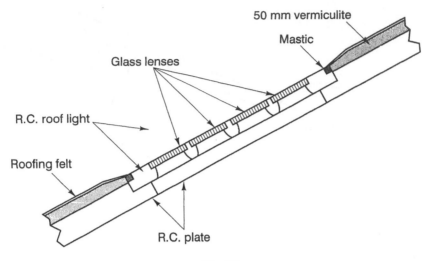

Fig. 3.3

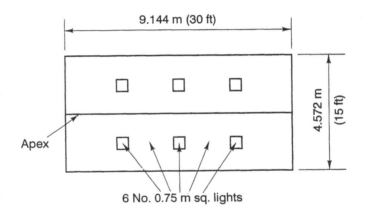

Fig. 3.4

The surrounds to the openings need to be reinforced so as to make up for the longitudinal and transverse reinforcement discontinued by the openings but required by the structural analysis and otherwise, as one would reinforce the surrounds of an opening made in a two-way spanning slab. Nominal corner bars are useful to resist local stresses at corners, as shown in photoelastic tests, but are not evaluated in analyses.

Ventilation can be provided by various types of ventilators using modest holes to accommodate them in the plates. Alternatively, some glazing can incorporate ventilation devices, or ventilation is arranged through the walls below, or in the gables.

Electrical conduits for lighting or operating fans or ventilators can be fastened to the soffits of the plates after construction. Alternatively, they can be cast in the plates, or bedded in, say, a 50 mm (2 in) vermiculite or a 25 mm (1 in) thick cork board insulation on top of the plates, and brought through the plate where required to either a fitment or a junction box.

Design tables are given in the Appendices of this book, and each Appendix commences with a figure showing the type of folded plate to which its tables refer.

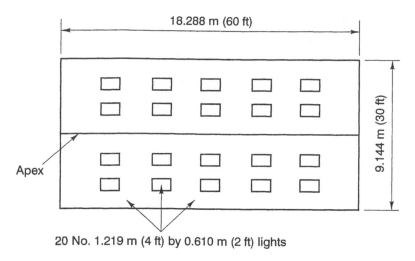

20 No. 1.219 m (4 ft) by 0.610 m (2 ft) lights

Fig. 3.5

3.2   Loading

First, there is the loading due to the self weight of the reinforced concrete folded plate roof and its various thickenings as shown on the drawings. This loading is included in the calculations for the tables of this book using a density for reinforced concrete of 24 kN/m³ (152.8 lb/ft³).

There are also other dead loads due to finishes to consider and there is the live snow load. With regard to finishes, the author has used – and believes that most shells (including folded plates) in the British Isles have – three layers of built-up roofing felt (the top one being mineral finished) on top of either 25 mm (1 in) thick cork board or 50 mm (2 in) thick vermiculite lightweight concrete. The soffit (i.e. the underside) of the shell or folded plate can be painted or plastered then painted.

British practice, Ref. 3.1, has, for many years, taken snow load (which includes loading due to cleaning and repair) as 0.75 kPa (15.66 psf) measured on plan area for roofs inclined at less than 30° to the horizontal. This has been overcome by the Department of Environment Building Regulations of 1994, which require the folded plates used in the tables of this book to take snow load (which includes loading due to cleaning and repair) as 1.0 kPa (20.89 psf) multiplied by the plan area of the roof. In the tables of this book the plates are inclined at 28° to the horizontal, so the equivalent snow load is 0.883 kPa (18.44 psf) multiplied by the sloping area of the roof. Consequently, this value can be added to the self weight of the plate and its finishes.

For North America, roofs experiencing the most severe snow loads are in Canada. The author is indebted to Professor John Christian (previously of the Universities of Bradford and Newfoundland) for the information which follows on snow loads on roofs in Canada. The National Building Code of Canada, 1975, gives a map specifying snow loads on the ground as basic pressures to which the design of roofs should be related in various locations of Canada. Extracts from this map are approximately as follows.

(a)   Calgary 20 psf (0.958 kPa) increasing to 120 psf (5.75 kPa) towards the Rocky Mountains.
(b)   Ottawa 60 psf (2.87 kPa).
(c)   Quebec 84 psf (4.02 kPa).
(d)   St John's, Newfoundland, 30 to 40 psf (1.44 to 1.92 kPa) increasing to 120 psf (5.75 kPa) towards North Labrador.
(e)   Toronto 40 psf (1.92 kPa).

(f)   Montreal 54 psf (2.59 kPa).
(g)   Northern and Eastern parts of Canada generally up to about 100 psf (4.79 kPa).
(h)   Generally less towards the USA, but still as great as 80 psf (3.83 kPa) at the USA/Canada border in the Rocky Mountains.

In the USA there is, for example, 80 psf (3.83 kPa) in the Rocky Mountains next to Canada, but also there are parts of the country that never experience snow.

As regards the snow loads for which roofs should be designed, these would normally be 0.8 times the above values of 'ground snow loads' specified for North America, according to the National Building Code of Canada.

For the USA, Ref. 3.2 incorporates a map, also given in Ref. 3.3, of the USA giving 'snow loads in pounds per square foot on the ground, 50 year mean recurrence interval'. This map excludes zones that, seemingly, include the Rocky Mountains, saying that these areas must base their snow loads on analyses of local climate and topography. Otherwise, the map gives zones ranging from 5 to 80 psf (0.24 to 3.83 kPa). These are useful if local codes, preferably based on statistical studies over an extended period, are not available. However, a minimum loading of 20 psf (0.958 kPa) is specified to provide for snow and construction and repair loads.

Certain USA books take snow load (which includes loading due to cleaning and repair) for folded plate roofs as 25 psf (1.197 kPa) of plan area, which is 1.057 kPa (22.08 psf) of sloping area of the roof.

For the UK the following loading of sloping areas was considered:

Superimposed (i.e. snow and maintenance)	0.883 kPa (18.4 psf)
	(i.e. 1.0 kPa (20.89 psf) per plan area)
Roofing felt	0.072
50 mm vermiculite	0.290
Plaster and paint	0.093
	1.338 kPa (27.9 psf)

For the USA the following loading of sloping area was considered.

Superimposed (i.e. snow and maintenance)	1.057 kPa (22.1 psf)
	(i.e. 25 psf (1.197 kPa) per plan area)
Roofing felt	0.072
25 mm cork board	0.060
Plaster and paint	0.093
	1.282 kPa (26.8 psf)

It was considered that a useful loading for finishes, snow and maintenance to use in the tables of this book is therefore 1.338 kPa (27.94 psf) multiplied by the sloping area of the roof. This means that, for the USA, a possible loading of sloping area is, for example, as follows.

Superimposed (i.e. snow and maintenance)	1.113 kPa (23.2 psf)
	(i.e. 26.34 psf (1.261 kPa) per plan area)
Roofing felt	0.072 kPa (1.5 psf)
25 mm cork board	0.060 kPa (1.3 psf)
Plaster and paint	0.093 kPa (1.9 psf)
	1.338 kPa (27.9 psf)

Alternatively, say, a possible loading of sloping area is as follows.

Superimposed (i.e. snow and maintenance)	1.206 kPa (25.1 psf)
	(i.e. 28.53 psf (1.366 kPa) per plan area)
Roofing felt	0.072 kPa (1.5 psf)
25 mm cork board	0.060 kPa (1.3 psf)
Paint	0.000
	1.338 kPa (27.9 psf)

Concerning loading due to wind, snow and maintenance the British Code of Practice, Ref. 3.1, has been replaced by Ref. 3.4 which says that Ref. 3.1 can still be used as an alternative.

Taking a fairly worst case for the UK (namely Scotland) for a typical folded plate roof with the plates inclined at 28° to the horizontal (i.e. as per the design tables in this book) the author calculated using Ref. 3.1 a maximum suction of 2.306 kPa (48.2 psf). This will not overcome the weight of, say, a 100 mm (4 in) plate weighing 2.36 kPa (49.3 psf) plus the weight of the finishes. The maximum wind pressure on a plate was 0.384 kPa (8.02 psf). If it is considered that a strong wind of this kind would remove any snow on the roof then the snow loading considered by the design should be more than satisfactory to allow for this wind pressure. In addition, the snow load in the design is not only greater but is on the whole roof, whereas the wind pressure is only on the windward slopes, with the leeward slopes simultaneously having suction. The total snow load for which the structure is designed is therefore far in excess of the possible downwards wind loading and one would therefore imagine the tables to be conservative in this respect if the wind is ignored. The designer may use the tables of this book and then check with his or her local Codes of Practice for wind suctions and pressures and, if he or she considers it necessary, because of the non-uniformity or excessive amount of these wind suctions and pressures, make an independent analysis for their effect.

The only way the writer can imagine that a depth of snow of about 813 mm (2 ft 9 in) could stay on a sloping roof if there were a powerful wind is for the sun to have melted the surface and then the surface to have frozen into an ice sheet to protect the snow from blowing away. While this has never been experienced by the writer (who has experienced considerable snow in his lifetime), the problem can be assessed by the designer for the location of the structure. The uniform snow load assumed by the code (Ref. 3.1) is unusual (Refs 3.5 and 3.4) if there is any wind at all when it is snowing, as wind causes the valleys to fill up and the ridges to be bare of snow.

The use of the code Ref. 3.4 is very complex for doubly pitched roofs as it considers various small parts of the roof to have different snow and wind loadings. Fortunately, the code Ref. 3.1, where the loading is uniform on each slope of a roof, can still be used.

Snow, maintenance and maximum wind loads are infrequent and exceedingly unlikely to occur simultaneously. Codes of Practice often make allowances for the latter and also allow for the fact that a short duration load is not as critical as a constant load, which, for example, is always there to be carried and can cause long range creep.

Of course, in the USA, hurricanes and tornados are sometimes experienced. Fortunately the latter are of low strength in the UK. Internationally, all sorts of wind and snow loadings occur and are dealt with by local Codes of Practice.

It should be noted that, if any loading that can be considered as uniformly spread is greater than that used by the tables in this book, then the tables can still be used to give a very approximate idea of the stresses and moments in the plates by increasing those in the tables by the ratio of the new total loading to the total loading used for the tables.

3.3   Buckling of plates

The buckling of plates is a topic that seems to have been ignored by previous works. A folded plate will commonly span a considerable distance relative to its thickness and its top portion will be in compression. For example, a plate 21.85 m (71.68 ft) long and 102 mm (4 in) thick has a span-to-thickness ratio of 214 to 1. This compares unfavourably with ACI and British Codes of Practice requiring lateral supports not to allow a span-to-thickness ratio greater than 50:1 and 60:1 respectively. Admittedly, one can say that the compression zone of a folded plate is like a column with a continuous side support, as the main compression zone is near the top part of a plate and this is restricted along its whole length by the next plate at the top. Also, reinforcement is designed to resist transverse moments. So, seemingly, the only concern is web buckling due to shear.

In the case of a plate supported on four sides and submitted to the action of shearing stresses uniformly distributed along the sides, using Equation (194) on p. 227 of Ref. 3.6, taking Poisson's ratio as 0.167 and L/h as greater than 3, the critical value of the shearing stress which may produce buckling of the plate is

$$4.568 \, E_c \, (t/h)^2 \tag{3.1}$$

This should be a reasonably conservative assessment compared with subsequent complex analyses. For example, if Young's modulus is 13 790 MPa (2 000 000 psi), $t = 102$ mm (4 in) and the critical value of the shearing stress which may produce buckling of the plate is limited to, say, 0.4827 MPa (70 psi), h is limited to 36.7 m (120.4 ft).

In producing the tables, the maximum shear stress is assessed for each plate and checked that it is satisfactory by Equation (3.1). Again, this is conservative as this maximum shear stress does not occur along all four sides and the sides are more fixed than free, that is, the plate is more rigidly fixed at its supports than is assumed by the Timoshenko formula, Equation (3.1).

When there is an upstanding beam or a cantilever at the edge, the top extremity may need lateral support by struts of 100 mm (or 4 in) cross-section, as shown in Figs 1.10 and 3.6 at centres, according to the code of practice one is using for designing the compression zone of a long slender rectangular beam. These struts could have four 10 mm (3/8 in) diameter bars, one in each corner, with 20 mm (3/4 in) cover and nominal 6 mm (1/4 in) stirrups (links) at 300 mm (12 in) centres.

For the spacing of these struts guided by ACI, Ref. 3.7, one might use:

$$50 \, t \tag{3.2}$$

Guided by Ref. 3.8, one might use the lesser of

$$60 \, t \quad \text{or} \quad 250 \, t^2/h \tag{3.3}$$

3.4   Thickness of plates

A great many singly and doubly curved shells have been constructed in the UK with shells only 64 mm (2.5 in) thick. There seems no reason therefore why folded plates could not be constructed as thin as this, bearing in mind that self weight is a large proportion of the total weight to be carried. The author has designed and constructed many 64 mm (2.5 in) thick shells and therefore does not subscribe to the view of Billington, Ref. 3.9, that a plate thinner than 102 mm (4 in),

Plate 1 Folded plate roofs for a teacher training college, Leeds, Yorkshire, UK. Each of the vaults shown is 3.35 m (11 ft) wide by 19.84 m (65 ft) span, whilst similar vaults over the gymnasium nearby are of 21.85 m (71 ft 8 in) span. The plates are 102 mm (4 in) thick. The columns are held together at their heads, at the gables, by 229 mm (9 in) wide by 152 mm (6 in) deep post-tensioned prestressed concrete ties. These were designed by the author.

Plate 2 Turin Exhibition Hall, Italy, designed by P.L. Nervi.

Plate 3 Sears Store, Tampa, Florida, USA (courtesy of Portland Cement Association, USA).

Plate 4 Air terminal, Minneapolis, USA (courtesy of Portland Cement Association, USA).

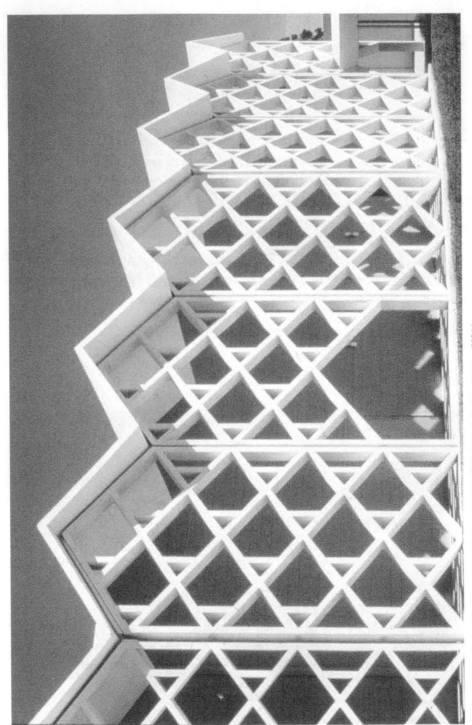

Plate 5 School in Albuquerque, USA (courtesy of Portland Cement Association, USA).

Plate 6 Swimming pool, Nottingham, UK (courtesy of Charles A.Wilby).

Plate 7 Entrance to Bradford College, Yorkshire, UK.

Plate 8 High School, Scottsdale, Arizona, USA (courtesy of Portland Cement Association, USA).

Plate 9 Bank in Colorado, USA (courtesy of Portland Cement Association, USA).

Plate 10 Chemical works, ICI Hyde, Cheshire, UK. These are reinforced concrete barrel vault roofs but show precast concrete lights comprising glass lenses framed in reinforced concrete which can be similarly used for folded plates. The wall lighting was similar so that no painting maintenance was envisaged as this was difficult to organise in the works' very short holiday.

Plate 11 These are the folded plates to the gymnasium mentioned in the caption to Plate 1. They show roof lighting similar to that shown in Plate 10 and similar end gable wall lighting.

Plate 12 Restaurant, Novato, California, USA (courtesy of Mark Stainburn Wilby).

Plate 13 St Paul's Church, Lorrimore Square, Walworth, London, UK (courtesy of British Cement Association, Crowthorne, Berkshire, UK).

Plate 14 Nuestra Señora del Rosario High School, Spain (courtesy of British Cement Association, Crowthorne, Berkshire, UK).

Plate 15 IBM Pavilion, Brussels Exhibition, Belgium (courtesy of British Cement Association, Crowthorne, Berkshire, UK).

Plate 16 American Concrete Institute Office, Detroit, Michigan, USA. The folded plates were 114 mm (4.5 in) thick (courtesy of British Cement Association, Crowthorne, Berkshire, UK).

Plate 17 Sonoma Development Centre, Eldridge, California, USA (courtesy of Mark Stainburn Wilby).

Plate 18 Coventry Pedestrian Precinct, UK

even though strong enough, should not be used because: 'From a practical standpoint such a thin slab is difficult and thereby costly to cast' – even though Billington designs a cylindrical shell only 75 mm (3 in) thick. However, for folded plates, 102 mm (4 in) thick seems to be a popular minimum thickness, although Ramaswamy, Ref. 3.10, designs one with 89 mm (3.5 in) thick plates.

Of course, the transverse moments are greater for folded plates than for shells. Shells are thickened, Refs 3.11 and 3.12, towards their valleys and edge beams to strengthen against transverse moments there, whereas folded plates are not similarly thickened. Also, curvature gives a membrane propping action against loading whereas plates do not. For these reasons, folded plates may need to be thicker than shells.

For plates where $L > 3h$, each plate is a long thin beam longitudinally, and essentially transversely spans between each fold. With regard to the latter transverse action, plates are continuous and, to limit deflection, the ACI code Ref. 3.7 states $t > h/28$, whereas British practice for many years was Table 7.1 of Ref. 3.13 which states $t > h/35$. Of course, deflection also depends on loading, which is unfortunately ignored by these recommendations. However, roof slabs resist very low loadings relative to other slabs, so these recommendations should be conservative. Involving loading gives the following.

From Table 7.2 of Ref. 3.13, the maximum deflection (because both ends are not fully fixed, assume one is fixed and the other free) is $wh^4/(185EI)$. Equating this to, say, $h/180$ (Table 9.5b of Ref. 3.7) and putting $I = t^3/12$ and allowing for shrinkage and creep, $E = 14\ 000\ 000$ kPa (~2 000 000 psi) gives $(h/t)^3 = 1\ 199\ 000/w$.

Assuming, for example, a roof slope of 28° and a live load of 0.9576 kPa (20 psf) gives $w = 0.9576 \times 0.8829 = 0.829$ kPa (17.32 psf) and therefore $h/t = 112.3$. For various values of t the corresponding maximum values of h and the corresponding values of L, assuming $L = 3h$, are given in Table 3.1.

Table 3.1

t (mm)	64	75	100	125	150
h (m)	7.19	8.42	11.23	14.04	16.85
L (m)	21.57	25.26	33.69	42.12	50.55
t (in)	2.5	3.0	4.0	5.0	6.0
h (ft)	23.40	28.08	37.43	46.79	56.15
L (ft)	70.20	84.24	112.29	140.37	168.45

However, Ramaswamy, Ref. 3.10, gives examples of 18.28 m (60 ft) span plates with thicknesses of 89 mm (3.5 in) and 102 mm (4 in), Billington, Ref. 3.9, gives examples of 10.67 m, 21.34 m and 32.00 m (35, 70 and 105 ft) span plates with thicknesses of 102 mm (4 in), Simpson, Ref. 3.14, gives an example of 18.29 m (60 ft) span plates with thicknesses of 102 mm (4 in), and the author has designed plates of 19.84 and 21.85 m (65.09 and 71.68 ft) span with thicknesses of 102 mm (4 in).

When end plates are cantilevered from the internal continuous plates, in order to limit deflection the ACI code, Ref. 3.7, states $t > h/10$ whereas in British practice, for many years, it was $t > h/12$. This was then altered, see Table 7.1, p. 221, and Table 3.10, p. 401, of Ref. 3.13, to $t > h/7$.

Involving loading gives the following.

The maximum deflection of a cantilever $= wh^4/(8E_cI)$. Equating this to say $h/180$ and putting $I = t^3/12$ and allowing for shrinkage and creep, $E_c = 14\ 000\ 000$ kPa (~2 000 000 psi) gives $(h/t)^3 = 51\ 850/w$.

Assuming, for example, a roof slope of 28° and a live load of 0.9576 kPa (20 psf) gives w = 0.9576 × 0.8829 = 0.8455 kPa (17.66 psf) and, therefore, h/t = 39.43. For various values of h the corresponding minimum values of t are given in Table 3.2.

Table 3.2

h (m)	0.6	0.9	1.2	1.5	1.8	2.1
$t = h/39.69$ (mm)	152	228	304	380	457	533
$t = h/10$ (mm)	60	90	120	150	180	210
h (ft)	2	3	4	5	6	7
$t = h/39.69$ (in)	0.61	0.91	1.22	1.52	1.83	2.13
$t = h/10$ (in)	2.4	3.6	4.8	6.0	7.2	8.4

However, Ramaswamy, Ref. 3.10, gives examples of 0.619 and 0.238 m (2.031 and 0.781 ft) end plates with thicknesses of 89 mm (3.5 in) and 102 mm (4 in), Billington, Ref. 3.9, gives an example of 0.914 m (3 ft) end plates with thicknesses of 152 mm (6 in) and Simpson, Ref. 3.14, gives an example of 1.926 m (6.32 ft) cantilever end plates with thicknesses of 127 mm (5 in).

The transverse plate deflections are traditionally restricted by codes of practice because of fears of cracking plaster ceiling finishes. This is fine for plaster ceiling finishes, but with other finishes what harm would more deflection do? This is for the designer to decide!

The author has made practical conservative calculations of deflections, plate strengths and reinforcement requirements in deciding the various thicknesses used in the design tables in the Appendices. The objective was that the sizes obtained by the use of the design tables would be adequate for deflection, strength and detailing of reinforcement.

Tables 3.1 and 3.2 should not therefore be used without considerations of strength, detailing of reinforcement and buckling. Essentially, Tables 3.1 and 3.2 are just an intellectual exercise for evaluation by the reader; the author, for example, would not use t less than 100 mm (4 in).

Considering lateral instability of the plates acting as long narrow beams, the longitudinal compression zone of each beam is restrained from buckling laterally by the adjacent plate. The only buckling to worry about is the diagonal compression as considered in Section 3.3. End beams and end cantilever plates are prevented from buckling – refer to Section 3.3 and, for example, see Fig. 3.6.

3.5 Concrete and cover to reinforcement

In the UK in the 1950s a considerable number of shell roofs were constructed. The author was then one of the few designers of shells, and the group of companies he worked for claimed to design more shell roofs than any other. In those days concrete mixes were generally specified by dry volumes of cement to sand to gravel (for example, refer to BSCP114, Ref. 3.12, and, for the USA, Ref. 3.7). A mix of 1:1.5:3 (cube strength at 28 days = 25.5 MPa (3698 psi), USA cylinder strength at 28 days = 3188 psi (22 MPa)) was favoured. Sometimes, a mix of 1:2:4 (cube strength at 28 days = 21 MPa (3046 psi), USA cylinder strength at 28 days = 2550 psi (17.58 MPa)) was used, but then there was a return to 1:1.5:3 mixes, which was also favoured by the main rival company. The return to 1:1.5:3 was because a certain North-light shell (not designed by the author) started flattening out and needed rectification by external prestressing. However, on subsequent investigation, the trouble was found not to be related to using the weaker 1:2:4 mix of concrete. Generally, the stresses in shells are not greater than can be resisted by a 1:2:4 mix of concrete. However, for folded plates the transverse moments are greater than for shells and, as the plates are thicker and therefore heavier, the longitudinal stresses are also greater.

Strut

Fig. 3.6

Since then, the UK (and, for the USA, see Ref. 3.7) has tended to use specially designed mixes for structural concrete. A problem experienced with this practice has been, on occasions, that mixes have been designed with a very low cement content for economy, and when ready-mix concrete was used, due to competition between suppliers, the strength specified was achieved with a very low water:cement ratio, necessitating very high compaction. When the cement content is very low it needs a much more thorough mixing, which it does not always receive because of inadequate time in the mixer due to pressure on the speed of output. In addition, with a low cement concrete, the small concrete specimens made to indicate strength are often compacted much more thoroughly and comprehensively than can be achieved for much larger *in situ* concrete volumes on site. Also, cement needs to be available to coat the reinforcement to help resist corrosion. For these reasons the British Code of Practice in 1972 (BSCP110), Ref. 3.15, specified minimum cement contents for concrete mixes, 'to ensure durability under specified conditions of exposure.' These specifications for structural grades of concrete in the later code of 1985 (Table 3.4, BS8110), Ref. 3.13, varied between 275 and 400 kg/m³ (17.2 and 25.0 lb/ft³) for various conditions of exposure. As an excessive proportion of cement increases shrinkage and cracking, the code BS8110 limits the maximum cement content to 550 kg/m³ (34.3 lb/ft³).

For durability, BS8110 relates the concrete cover to the reinforcement to a minimum cement content between 275 and 400 kg/m³ (17.2 and 25.0 lb/ft³), a maximum water to cement ratio, to strength and to exposure condition.

As already described, there has been a wealth of experience in the UK with shells which were often 64 mm (2.5 in) thick. The concrete cover to the reinforcement was commonly 13 mm (0.5 in) and the outer reinforcement bars were not more than this in diameter. This was to maximise the effective depth of the reinforcement for transverse moments.

The basic requirements of cover are not to impair the bond/adhesion between the reinforcement and concrete for structural reasons, to provide protection against corrosion, and to protect for a specified time the steel from the weakening effects of a fire. The 12.7 mm (0.5 in) cover for bars with diameters less than or equal to 12.7 mm (0.5 in) proved adequate for the first two of these requirements. Fire regulations vary internationally but it was often the practice in the UK not to restrict roof design – permitting, for example, unclad structural steel portals, trusses, etc. to support roofs. So the 12.7 mm (0.5 in) cover was allowed by fire regulations.

In one example, Ref. 3.13, severe corrosion of the shell reinforcement occurred over about 30 years, even though the shell was well protected with roofing felt. This proved to be due to the presence of calcium chloride (not now used) as a hardener in the concrete mix.

When there are no detrimental chemicals, the above practice concerning concrete quality and cover has seemingly proved satisfactory for over about 40 years.

ACI 318-89, Ref. 3.7, requires that shells should use concrete with a cylinder compressive strength not less than 20.7 MPa (3000 psi), which is approximately a British cube strength of 24.3

MPa (3529 psi). As the same code of practice mentions, the value of Young's modulus increases for stronger concretes and so helps resistance to buckling.

BS 8110 and BS 5321 do not recommend the use of concretes with normal weight aggregates of characteristic strengths less than 25 MPa (3626 psi).

ACI 318-89 requires that for shells not exposed to the weather, the minimum concrete cover to the reinforcement should be 0.5 in (~ 13 mm) for a 0.625 in (~ 16 mm) bar and smaller, W31 or D31 wire, and 0.75 in (~ 19 mm) for a 0.75 in (19 mm) bar and larger.

BS 8110 for normal weight concretes states that, where the concrete surfaces are protected against weather or aggressive conditions, the normal maximum size of the aggregate should be no more than 15 mm (~ 0.625 in) and if a 'systematic checking regime is established to ensure compliance with the limits' of the water/cement ratio and cement content, the minimum concrete covers can be as given in Table 3.3.

Table 3.3

Concrete	Nominal cover (mm)				
	25	20	15	15	15
Max. free water/cement ratio	0.65	0.60	0.55	0.50	0.45
Min. cement content (kg/m³)	275	300	325	350	400
Min. characteristic compressive strength (MPa)	25	30	35	40	45

To summarise, the designer will be guided by his or her own experience and the codes of his or her country, and will consider the exposure conditions and protection given to the concrete, the latter perhaps ranging from nothing, to painting, to say three layers of built-up roofing felt in the UK. In the UK, the plate concrete would be protected from rainwater, and a 1:1.5:3 mix of cement:sand:gravel by dry volumes or dry weights, using normal aggregates, should be satisfactory. This mix has a cement content of 437 kg/m³ (27.3 lb/ft³) and needs to have a minimum cube strength at 28 days of 25.5 MPa (3698 psi) or a minimum USA cylinder strength of 3188 psi (22 MPa). It should be mixed thoroughly and should be wet enough, but not more so, to be compacted very well around the reinforcement by whatever means available. All of the concrete needs compacting thoroughly, remembering (Ref. 3.13) that 5, 10 and 25% of air voids can give losses in strength of 30, 60 and 90% respectively. If inadequate vibrators are present for this then the concrete needs to be wetter. It is not unusual in the UK for unpredictable rainfall to occur during concreting. For light rainfall the concrete may be made much drier to allow for this. For heavy rainfall the concrete has to be stopped off and covered. In a hot dry country an unexpected severe wind can dry out the concrete unduly as it is being placed, so the water content of the mix may need increasing.

In the UK the author has often decided upon the water content by taking a handful of concrete and throwing it away. When one's hand is left with specks of grey water, rather than complete dryness and is not completely coated with grey water, it is acceptable for vibrating with an internal (poker or sausage type) vibrator until the concrete surface is seen to 'swim'. It should be vibrated no further so as not to cause segregation.

Saving cement through the mix design requiring longer mixing time and better vibration makes a fairly negligible saving in the cost of a complete building contract, and if the better mixing and vibration is not consistently maintained the concrete will be condemned. Also, if longer mixing time and better vibration are conscientiously achieved it might be more expensive than using the 1:1.5:3 mix concrete described. Of course, to a ready-mix supply firm the cement cost is significant. Mix design can of course improve the ratio of fine to coarse aggregates in the 1:4.5 cement to aggregates mix.

With regard to the cover, the designer will be guided by his or her own experience and the codes of practice of his or her country, and will consider the exposure conditions and protection given to the concrete.

3.6 Reinforcement

It has been established practice over many decades to analyse framed structures elastically as though uncracked, and then to design reinforcement with working/permissible stresses such as 138 MPa (20 000 psi) for mild steel and 207 MPa (30 000 psi) for high-tensile deformed bars. This situation is also aggravated by shrinkage, and the total result means that the concrete must be cracked. For example if the reinforcement is stressed to 207 MPa (30 000 psi) the tensile stress in the concrete in contact with it, using a modular ratio of 15, would be 14 MPa (2000 psi), which would cause the concrete to crack. In addition, shrinkage would help cracking.

The author worked for a firm which was a pioneer of the use of deformed high-tensile reinforcement. When a certain bar was introduced, tests were made to show that the maximum crack size at working loads due to bending stresses for a beam were no greater for this particular deformed bar than for mild steel, because the former distributed the cracks better (refer to p. 29 and Ref. 2 of Ref. 3.12) due to the mechanical bond. This same company was a pioneer of shell roof design in the UK and used high-tensile steel fabrics top and bottom in shells to resist moments and high-tensile deformed bars to resist tensile forces (Ref. 3.12). The moments and forces had been obtained from an elastic analysis assuming uncracked sections and one was more worried about this illogicality for the thin shells than for the well established frames. However, like the frames the procedure proved satisfactory.

After leaving that company, the author decided, because of the anomaly just discussed, to use mild steel bars for subsequent shell designs, apart from keeping to the fabrics because these generally provided more steel than necessary and so were not fully stressed in all or most places. The fabrics gave a good distribution of reinforcement and were easy to place. In addition, the 'Spacer bars' (p. 187 of Ref. 3.12) were kept as the gross diameter of square twisted bars was more economic than for mild steel bars of circular cross-section and they spread any cracks better.

The tables of this book have similarly used elastic theory for uncracked sections and therefore, by the same argument, reinforcement suitable for general reinforced concrete design can seemingly be used for folded plates.

For parts of the folded plates the analysis may indicate that tensile reinforcement is not necessary. On these occasions a minimum amount of reinforcement is provided to resist temperature and shrinkage stresses and local stresses due, for example, to the activities of maintenance workers. BS 8110 recommends minimum reinforcement of 0.24% and 0.13% of the gross cross-sectional area for plain mild steel bars and high-yield high-bond bars or high-yield mesh respectively. ACI 318–89 requires these quantities to be 0.2 and 0.18 respectively.

3.7 Notes on tables

Span = L; breadth or width = B; plate thickness = t; F0 = longitudinal stress at edge No. 0; F1 = longitudinal stress at fold No. 1; F2 = longitudinal stress at fold No. 2; F3 = longitudinal stress at fold No. 3; M0 = 0; M1 = transverse moment at fold No. 1; M2 = transverse moment at fold No. 2; M3 = transverse moment at fold No. 3. Longitudinal stresses are positive when in tension. Moments are positive when sagging, i.e. endeavouring to induce tensile stresses in the bottom fibres. All distances are in metres. All stresses are in MPa. All moments are in kNm/m. The analysis for the tables has taken Poisson's ratio as zero, apart from the buckling check equation in Section 3.3.

Tables UUEB (3 folds) refers to a folded plate roof with Unpropped Upstand Edge Beams and three folds, namely two valleys next to the two edge beams and an apex, and $d = L/15$ or not less than 0.76 m. Refer to Figs A1 and A2. In Fig. A1 the plate numbering refers to this unpropped case. In Fig. A2 the plate numbering refers to the propped case of Tables PUEB.

Tables PUEB (3 folds) refers to a folded plate roof with Propped Upstand Edge Beams and three folds, namely two valleys next to the two edge beams and an apex, and $d = L/15$ or not less than 0.76 m. Refer to Figs A1 and A2. In Fig. A2 the plate numbering refers to this propped case. Note that when an end plate is propped, it is considered as a short wide horizontal plate from 0 to 1 points and F0 and F1 are therefore at each side of the edge beam.

Tables CUUEB (5 folds) refers to a Continuous folded plate roof with Unpropped Upstand Edge Beams and five folds, namely two valleys next to the two edge beams, one central valley and two apexes, and $d = L/15$ or not less than 0.76 m. Refer to Figs A3 and A4 (see also A2). In Fig. A3 the plate numbering refers to this unpropped case. In Fig. A2 the plate numbering refers to the propped case of Tables CPUEB.

Tables CPUEB (5 folds) refers to a Continuous folded plate roof with Propped Upstand Edge Beams and five folds, namely two valleys next to the two edge beams, one central valley and two apexes, and $d = L/15$ or not less than 0.76 m. Refer to Figs A3 and A4 (see also A2). In Fig. A2 the plate numbering refers to this propped case.

Tables UDEB (5 folds) refers to a folded plate roof with Unpropped Downstand Edge Beams and five folds. Refer to Figs A5 and A6. In Fig. A5 the plate numbering refers to this unpropped case. In Fig. A6 the plate numbering refers to the propped case of Tables PDEB.

Tables PDEB (5 folds) refers to a folded plate roof with Propped Downstand Edge Beams and five folds. Refer to Figs A5 and A6. In Fig. A6 the plate numbering refers to this propped case.

Tables CUDEB (9 folds) refers to a Continuous folded plate roof with Unpropped Downstand Edge Beams and nine folds. Refer to Figs A7 and A8. In Fig. A7(a) the plate numbering refers to this unpropped case. In Fig. A7(b) the plate numbering refers to the propped case of Tables CPDEB.

Tables CPDEB (9 folds) refers to a Continuous folded plate roof with Propped Downstand Edge Beams and nine folds. Refer to Figs A7 and A8. In Fig. A7(b) the plate numbering refers to this propped case.

Tables UEC (3 folds) refers to a folded plate roof with End Cantilevers and three folds, namely two Unpropped valleys and an apex. Refer to Figs A9 and A10.

Tables CUEC (5 folds) refers to a Continuous folded plate roof with End Cantilevers and five folds, namely three Unpropped valleys and two apexes. Refer to Figs A11 and A12.

References

3.1   BS CP3 (1972) Chapter 5, Loading.

3.2   USASI A-59.1 (1971) *Building Code Requirements for Minimum Design Loads in Buildings and Other Structures*, USA Standards Institute, New York.

3.3   Winter, G. and Nilson, A.H. (1972) *Design of Concrete Structures*, Ninth Edition, McGraw-Hill, New York, USA.

3.4   BS 6399 (1995).

3.5   Parkinson, J. (1982) Roof shape is key to snow load, *New Civil Engineer*, January.

3.6   Timoshenko, S.P. (1948) *Strength of Materials, Part 2. Advanced Theory and Problems*, D. Van Nostrand Co., New York and London.

3.7   ACI 318-89, ACI 318R-89 (1989).

3.8   BS8110 (1985).

3.9   Billington, D.P. (1965) *Thin Concrete Shell Structures*, McGraw-Hill, New York, USA.

3.10 Ramaswamy, G.S. (1984) *Design and Construction of Concrete Shell Roofs*, Krieger, Malabar, Florida, USA.

3.11 Wilby, C.B. and Khwaja, I. (1977) *Concrete Shell Roofs*, Applied Science Publishers, London; Wilby, C.B. (1980) *Design Graphs for Concrete Shell Roofs*, Applied Science Publishers, London; and Ref. 3.12.

3.12 Evans, R.H. and Wilby, C.B. (1963) *Concrete: Plain, Reinforced, Prestressed and Shell*, Edward Arnold, London.

3.13 Wilby, C.B. (1983) *Concrete Materials and Structures*, Cambridge University Press, Cambridge, UK, and New York, USA.

3.14 Simpson, H. (1958) Design of folded plate roofs, *Proc Am Soc Civil Engrs*, January.

3.15 Wilby, C.B. (1983) *Structural Concrete*, Butterworths, London, UK, and Boston, USA.

4

Examples of the use of design tables

4.1 Introduction

Shell roofs have been satisfactorily designed using the elastic design of BSCP114 over many years, Ref. 4.1. Low reinforcement stresses agree better with the elastic analyses used for producing the design tables of this book. In the following examples, elastic analysis is therefore used along with mild steel reinforcement. This practice is favoured by the author. Elastic analysis is described in Refs 4.1, 4.2, 4.3 and 4.4. Permissible stresses for concrete and steel are given in BSCP114, Metric Units, 1969.

The modular ratio = 15.

The reader should study all the examples in this chapter before commencing a design, to avoid repetition, as on occasions later examples rely on the reader having studied earlier examples.

4.2 Type UUEB (as shown in Figs A1 and A2)

Data: $L = 19$ m (62.34 ft), for loading see Section 3.2 (1.338 kPa (27.94 psf) of sloping area plus self weight), $B = 10.21$ m (35.5 ft), $t = 0.14$ m (5.51 in).

This lies between Tables UUEB 112 and 118.

Using linear interpolation:

$$d = 1.25 + (1.28 - 1.25) \times [(19.0 - 18.69)/(19.25 - 18.69)]$$
$$= 1.25 + (1.28 - 1.25) \times [0.5536] = 1.267 \text{ m } (4.157 \text{ ft})$$

$$F0 = -9.33 - (-9.56 + 9.33) \times [0.5536] = -9.457 \text{ MPa } (1372 \text{ psi})$$

$$F1 = 5.14 + (5.35 - 5.14) \times [0.5536] = 5.256 \text{ MPa } (762.3 \text{ psi})$$

$$F2 = -\{3.64 + (3.82 - 3.64) \times [0.5536]\} = -3.74 \text{ MPa } (542.4 \text{ psi})$$

$$M1 = 0$$

$$M2 = 24.41 + (24.78 - 24.41) \times [0.5536] = 24.61 \text{ kNm/m}$$

4.2.1 Edge beam

Using BSCP114, F0 requires a 1:1:2 mix concrete, permissible compressive stress = 10 MPa (1450 psi) (cube strength at 28 days = 30 MPa (4351 psi), about a USA cylinder strength of 26 MPa (3771 psi)). Then no compression reinforcement is required. The distribution of longitudinal stress is as shown in Fig. 4.1.

The height of the neutral axis = $5.256 \times 1.267/(5.256 + 9.457) = 0.4526$ m (17.82 in)

In Fig. 4.1 the longitudinal tension force

$$= 0.5 \times 5.256 \times 0.4526 \times 0.23 = 0.2736 \text{ MN}$$

The permissible tensile stress in the steel = 140 MPa.

The area of steel required = $0.2736/140$ m² = 1954 mm².

Use (from Table 3.2 of Ref. 4.4) four bars of 25 mm diameter (1 in diameter).

On the one hand, it can be thought that the centroid of these bars should correspond to the centroid of the triangular stress block, that is, at a height of $0.4526/3 = 0.1509$ m. On the other hand, one normally puts the tensile steel as low as possible in a beam. The author tends to do and recommend the latter, but spreads the bars out fairly generously upwards. In this example, there are not many bars to demonstrate this, but using a cover to the bottom two bars of 25 mm (1 in) a vertical gap of 50 mm, say, can be put between these and the next layer of two bars.

The minimum area of reinforcement in a 230 mm thick member (Section 3.6) = $230 \times 1000 \times 0.24/100 = 552$ mm²/m or, per side, = $552/2 = 276$ mm²/m. Thus, use 8 mm diameter bars at 175 mm centres at each side of the member.

Maximum spacing of reinforcement allowed = $3 \times (230 - 25 - 4) = 603$ mm.

The stresses in Fig. 4.1 can be considered as being due to an axial longitudinal compression C and a moment M, where C causes a uniform stress = $(9.457 + 5.256)/2 - 5.256 = 7.357 - 5.256 = 2.101$ MPa, and M causes a bending stress = 7.357 MPa.

Figure 4.2 shows the stresses indicating this point.

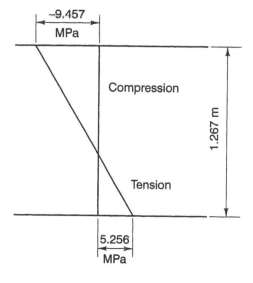

Fig. 4.1

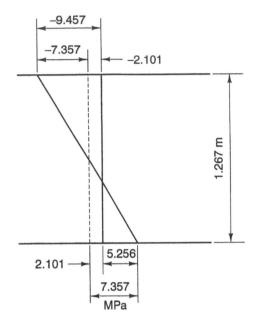

Fig. 4.2

Therefore,

$$C = 2.101 \times 0.23 \times 1.267 = 0.6123 \text{ MN}$$

and

$$M = 7.357 \times 0.23 \times 1.267^2/6 = 0.4528 \text{ MNm}$$

The longitudinal axial compressive stress of 2.101 MPa will have, say, a cosine distribution from the centre to the support (or a parabolic distribution – maximum at the centre and zero at the supports) so it will not relieve the diagonal tensile stress due to the maximum shear stress.

Considering $M = wl^2/8$, then

$$\text{maximum shear force} = wl/2 = 4M/l = 4 \times 0.4528/19 = 0.09533 \text{ MN}$$

Taking a parabolic distribution of shear stress, maximum shear stress = $1.5 \times 0.09533/(0.23 \times 1.267) = 0.4907$ MPa (71.17 psi). For a 1:1:2 mix concrete, BSCP114 recommends a permissible shear stress of 0.9 MPa (130.5 psi), but the author prefers the limits recommended in Ref. 4.1 of 0.7309 MPa (106 psi) except for rectangular beams with more than 0.17% of longitudinal compression reinforcement, when the limit should be 0.5447 MPa (79 psi). Whichever way, no shear reinforcement is required. Had shear reinforcement been required the treatment would have been similar to that in Section 4.2.2. However, nominal stirrups (links) must be used, say 10 mm diameter at 230 mm centres (or 0.375 in diameter at 9 in centres). This provides more than the minimum requirement of 552 mm²/m calculated previously.

Had the edge beam required compression reinforcement, this could be designed like a similar beam in Ref. 4.5.

Referring to Equations (3.2) and (3.3), 50 t = 50 × 0.14 = 7 m and 250 t/h = 250 × 0.14/5.782 (given later) = 6.053 m. Use three intermediate struts 100 mm × 100 mm (4 in × 4 in) similar to those shown in Fig. 1.10, i.e. at 19/4 = 4.75 m (15.58 ft) spacing.

4.2.2 Sloping plate

As mentioned previously, use a 1:1:2 mix concrete, as this is required for the edge beam. As F2 = − 3.74 MPa (542.4 psi), which is less than the permissible compressive strength of 10 MPa (1450 psi), no compression steel is required.

The distribution of longitudinal stresses is shown in Fig. 4.3.

$$h = B/(2 \cos 28°) = 10.21/(2 \cos 28°) = 5.782 \text{ m (18.97 ft)}$$

The height of the neutral axis = 5.256 × 5.782/(5.256 + 3.74)

$$= 5.256 × 0.6427 = 3.378 \text{ m (11.08 ft)}$$

From Fig. 4.3 the longitudinal tension force can be taken as

$$= \text{F1} × 3.378 × t/2 = 5.256 × 3.378 × 0.14/2 = 1.243 \text{ MN}$$

The area of tensile steel = 1.243/140 m² = 8879 mm².

Use 18 bars of 25 mm diameter and 1 bar of 8 mm diameter (18 bars of 1 in diameter and 1 bar of 0.375 in diameter).

These can be arranged so that their centroid corresponds to the centroid of the triangle of tensile stress. Alternatively, while the author tends to bear this in mind to some extent, he concentrates the bars more towards the bottom, as one does in a beam. He suggests using 100 mm

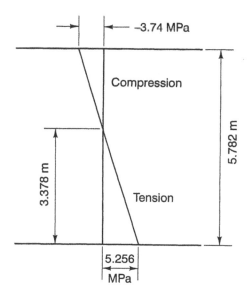

Fig. 4.3

spacing so that the whole group covers a distance = 18 × 0.1 = 1.8 m (5.906 ft). These bars shall have a cover of at least 25 mm (1 in) and can be placed anywhere in the depth of the plate and are usually placed alternately on top of the bottom steel for bending and beneath the top of the top steel for bending.

Minimum area of steel in 0.14 m, 5.51 in, thick plate

$$= (0.24/100) \times 140 \times 1000 = 336 \text{ mm}^2/\text{m}$$

Sharing this reinforcement between top and bottom, each has 336/2 = 168 mm²/m.

Use 8 mm diameter bars at 300 mm centres, or 0.375 in diameter bars at 12 in centres, both top and bottom.

$$\text{Maximum spacing allowed} = 3 \times (140 - 25 - 4) = 333 \text{ mm}$$

Therefore, from the bottom, after the 1.8 m distance accommodating the tensile steel, use 8 mm diameter bars at 300 mm centres top and bottom, beneath the top and above the bottom transverse bending reinforcement respectively, to act as its distribution steel.

The stresses in Fig. 4.3 can be considered as due to an axial longitudinal tension T and moment M where T causes a uniform stress

$$= (3.74 + 5.256)/2 - 3.74 = 0.758 \text{ MPa}$$

M causes a bending stress = 4.498 MPa.

Figure 4.4 shows the Fig. 4.3 stresses indicating this point. Therefore

$$T = 0.758 \times 0.14 \times 5.782 = 0.6136 \text{ MN}$$

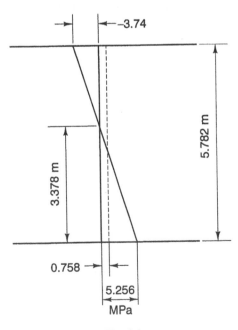

Fig. 4.4

This has a cosine or parabolic distribution from the maximum at mid-span to zero at the supports and

$$M = 4.498 \times 0.14 \times 5.782^2/6 = 3.509 \text{ MN/m}$$

As previously

$$\text{maximum shear force} = 4M/l = 4 \times 3.509/19 = 0.7387 \text{ MN}$$

then taking a parabolic distribution of shear stress

$$\text{maximum shear stress} = 1.5 \times (0.7387/0.14)/5.782 = 1.369 \text{ MPa (199 psi)}$$

As T is zero at the supports, the maximum principal tensile stress at the supports is the maximum shear stress, i.e. 1.369 MPa (199 psi), so diagonal tension reinforcement is required and = $1.369 \times 0.14/140$ m²/m = 1369 mm²/m.

Use say 16 mm diameter bars at 150 mm centres, or 5/8 in diameter bars at 6 in centres. These are at 45° to the longitudinal axis. The spacing can be increased towards the apex and valley in accordance with the parabolic distribution of shear stress. Also, the diagonal tension reinforcement can be reduced towards mid-span. For example, for a section 2 m from the support, the shear stress = $[2/(19/2)] \times 1.369 = 0.2882$ MPa and the longitudinal tensile stress = $0.758 \times [(9.5 - 2)/9.5]^2 = 0.4724$ MPa.

The maximum principal tensile stress resulting from these stresses

$$= 0.5 \times (0.4724 + (0.4724^2 + 4 \times 0.2882^2)^{0.5})$$
$$= 0.6088 \text{ MPa, } 88.30 \text{ psi}$$

From Section 4.2.1 one would perhaps not use diagonal tension steel. For the purpose of illustration assume tension steel is required.

The angle of this principal stress is given by

$$\tan 2\theta = 2 \times 0.2882/0.4724 = 1.220$$
$$\text{i.e. } \theta = 25.33°$$

Use reinforcement at 45° to the longitudinal axis to keep the same system as at the supports.

$$\text{Diagonal reinforcement required} = 0.6088 \times 0.14/140 \text{ m}^2/\text{m} = 608.8 \text{ mm}^2/\text{m}$$

This is required at 25.33°, so putting it at 45° is less effective; consequently, the reinforcement required at 45° is

$$= 608.8/\cos (45 - 25.33) = 646.5 \text{ mm}^2/\text{m}$$

Use 16 mm diameter bars at 300 mm centres or 5/8 in diameter bars at 12 in centres.

With regard to the transverse bending moments: M1 = 0 and M2 = 24.61 kNm/m.

The loading = 1.338 kPa (27.94 psf) (see Section 3.2) of the sloping area plus self weight, which is $0.14 \times 24 = 3.360$ kPa (70.17 psf) = 4.698 kPa (98.12 psf) of the sloping area; that is, 5.321 kPa (111.1 psf) of the plan area.

The maximum mid-span bending moment for free supports at a valley and ridge = $[4.698 \times 5.782 \times 10.21/2]/8 = 17.33$ kNm/m.

Figure 4.5 shows the transverse bending moment diagram. At the apex using BSCP114 elastic design, the permissible concrete compressive stress in bending = 10 MPa (1450 psi) and the

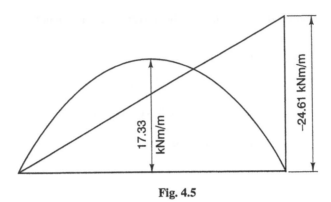

Fig. 4.5

permissible tensile stress in steel = 140 MPa (20 300 psi), and using the formulas from p. 95 of Ref. 4.4:

$\alpha_f = 140/10 = 14$
$\alpha_e = 15$
$x_1 = 15/(15 + 14) = 0.5172$
$z_1 = 1 - x_1/3 = 0.8276$
$K = 0.5 \times 10 \times 0.5172 \times 0.8276 = 2.140$

Using 20 mm diameter bars and 20 mm cover, the effective depth of reinforcement = 0.14 – 0.03 = 0.11 m.

moment of resistance = $2.14 \times 1 \times 0.11^2 = 0.2589$ MNm/m

which is greater than 24.61 kNm/m = 0.02461 MNm/m.
The area of tensile reinforcement

= $0.02461/(0.8276 \times 0.11 \times 140) = 0.001931$ m²/m
= 1931 mm²/m

Use 20 mm diameter bars at 150 mm centres (or 7/8 in diameter bars at 6 in centres).
   The bending moment at a horizontal distance X from the edge beam

= $5.321 \times (5.105/2) \times X - 5.321 \times X^2/2 - 24.61 \times X/5.105$

By calculus, a maximum occurs at $X = 1.647$ m and has the value = 7.213 kNm/m = 0.007213 MPa.
   The maximum area of tensile reinforcement required in the bottom of the plate

= $0.007213/(0.8276 \times 0.11 \times 140)$ m²/m = 565.9 mm²/m

Use 12 mm diameter bars at 200 mm centres (or 1/2 in diameter bars at 8 in centres). These can be reduced towards the ridge and valley but must not reduce to less than the minimum requirements: from previously, this can be taken as 168 mm²/m.
   Figure 4.6 indicates the reinforcement system at mid-span (not to scale and not showing every bar).
   The longitudinal steel in the edge beam and sloping plate required structurally at mid-span can be reduced towards the supports, similarly to a simply supported beam carrying a uniformly

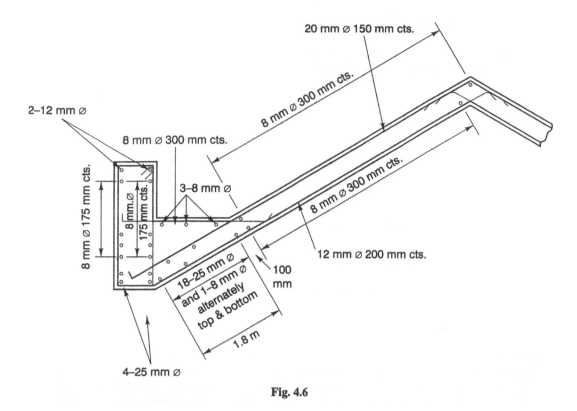

Fig. 4.6

distributed load. The minimum reinforcement used but not required structurally cannot be reduced.

The author introduced the teaching of folded plate roofs to both final year Batchelor and Master degree students at the University of Bradford. This was continued by Messrs D. Walker and R. Westbrook. The latter used the author's work on this subject as a basis for the chapter on folded plates in his book, Ref. 4.6, and readers might find his book of interest.

4.3 Type CUUEB (as shown in Figs A3 and A4; see also A2)

Data: $L = 19$ m (62.34 ft), for loading see Section 3.2 (1.338 kPa (27.94 psf) of sloping area plus self weight).

This lies between Tables CUUEB 112 and 118 (but can lie between 111 and 117, etc., according to which B is preferred).

$B = 10.21$ m (35.5 ft), $t = 0.14$ m (5.51 in)

Using linear interpolation:

d = as in Section 4.2 = 1.267 m (4.157 ft)
$F0 = -9.69 - (9.93 - 9.69) \times (19 - 18.69)/(19.25 - 18.69)$
$\quad = -9.69 - (9.93 - 9.69) \times [0.55357] = -9.823$ MPa $(-1425$ psi)
$F1 = 5.16 + (5.39 - 5.16) \times [0.55357] = 5.287$ MPa $(766.8$ psi)
$F2 = -3.55 - (3.74 - 3.55) \times [0.55357] = -3.655$ MPa $(-530.1$ psi)

F3 = 3.55 + (3.76 − 3.55) × [0.55357] = 3.666 MPa (531.7 psi)
M1 = 0
M2 = − 18.97 − (19.19 − 18.97) × [0.55357] = − 19.09 kNm/m = − 4292 lb ft/ft
M3 = − 8.03 + (8.03 − 7.94) × [0.55357] = − 7.98 kNm/m = − 1794 lb ft/ft

4.3.1 Edge beam

Using BSCP114, F0 requires a 1:1:2 mix concrete, permissible compressive stress = 10 MPa (1450 psi) (cube strength at 28 days = 30 MPa (4351 psi), about a USA cylinder strength of 26 MPa (3771 psi)). Then no compression reinforcement is required. The distribution of longitudinal stress is as shown in Fig. 4.7.

The height of the neutral axis

$$= 5.287 \times 1.267/(5.287 + 9.823) = 0.4433 \text{ m } (17.45 \text{ in})$$

In Fig. 4.7 the longitudinal tension force

$$= 0.5 \times 5.287 \times 0.4433 \times 0.23 = 0.2695 \text{ MN}$$

The permissible tensile stress in the steel = 140 MPa.

The area of steel required = 0.2695/140 m² = 1925 mm².

Use (from Table 3.2 of Ref. 4.4) four bars of 25 mm diameter (1 in diameter).

On the one hand it can be thought that the centroid of these bars should correspond to the centroid of the triangular stress block; that is, at a height of 0.4433/3 = 0.1478 m. On the other hand, one normally puts the tensile steel as low as possible in a beam. The author tends to do, and recommend, the latter, but spreads the bars out fairly generously upwards. There are not many bars to demonstrate this in this example, but using a cover to the bottom two bars of 25 mm (1 in) a vertical gap of 50 mm, say, can be put between these and the next layer of two bars.

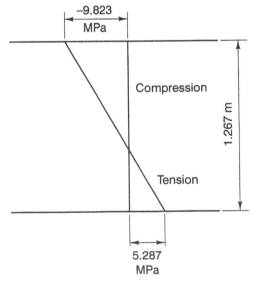

Fig. 4.7

The minimum area of reinforcement in a 230 mm thick member (Section 3.6) = 230 × 1000 × 0.24/100 = 552 mm²/m or per side = 552/2 = 276 mm²/m. Thus, use 8 mm diameter bars at 175 mm centres at each side of the member.

Maximum spacing of reinforcement allowed = 3 × (230 − 25 − 4) = 603 mm

The stresses in Fig. 4.7 can be considered as due to an axial longitudinal compression C and a moment M where C causes a uniform stress = (9.823 + 5.287)/2 − 5.287 = 7.555 − 5.287 = 2.268 MPa and M causes a bending stress = 7.555 MPa.

Figure 4.8 shows the stresses indicating this point. Therefore

$$C = 2.268 \times 0.23 \times 1.267 = 0.6609 \text{ MN}$$

and

$$M = 7.555 \times 0.23 \times 1.267^2/6 = 0.4649 \text{ MNm}$$

The longitudinal axial compressive stress of 2.268 MPa will have, say, a cosine distribution from the centre to the support (or a parabolic distribution, with the maximum at the centre and zero at the supports) so it will not relieve the diagonal tensile stress due to the maximum shear stress at the supports.

Considering $M = wl^2/8$ then

$$\text{maximum shear force} = wl/2 = 4M/l = 4 \times 0.4649/19 = 0.09787 \text{ MN}$$

Taking a parabolic distribution of shear stress, maximum shear stress = 1.5 × 0.09787/(0.23 × 1.267) = 0.5038 MPa (73.07 psi). For a 1:1:2 mix concrete, BSCP114 recommends a permissible

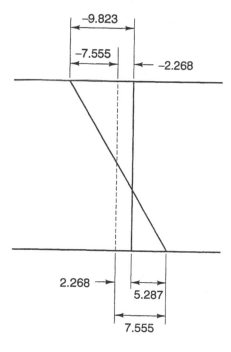

Fig. 4.8

shear stress of 0.9 MPa (130.5 psi), but the author prefers the limits recommended in Ref. 4.1 of 0.7309 MPa (106 psi) except for rectangular beams with more than 0.17% of longitudinal compression reinforcement, when the limit should be 0.5447 MPa (79 psi). Whichever limits are used, no shear reinforcement is required. Had shear reinforcement been required the treatment would have been similar to that of Section 4.3.2. However, nominal stirrups (links) must be used, say 10 mm diameter at 230 mm centres (or 0.375 in diameter at 9 in centres). This provides more than the minimum requirement of 552 mm²/m calculated previously.

Had the edge beam required compression reinforcement, this could be designed like a beam similar to that in Ref.4.5.

The same struts as designed in Section 4.2.1 are required to prevent lateral instability of this edge beam.

4.3.2 Sloping plate 1–2

As mentioned previously, use a 1:1:2 mix of concrete, as this is required for the edge beam. As $F2 = -3.655$ MPa (530.1 psi), which is less than the permissible compressive strength of 10 MPa (1450 psi), no compression steel is required.

The distribution of longitudinal stresses is shown in Fig. 4.9.

$$h = B/(2 \cos 28°) = 10.21/(2 \cos 28°) = 5.782 \text{ m } (18.97 \text{ ft})$$

The height of the neutral axis

$$= 5.287 \times 5.782/(5.287 + 3.655) = 5.287 \times 0.6466 = 3.419 \text{ m } (11.22 \text{ ft})$$

From Fig. 4.9 the longitudinal tension force can be taken as

$$= F1 \times 3.419 \times t/2 = 5.287 \times 3.419 \times 0.14/2 = 1.265 \text{ MN}$$

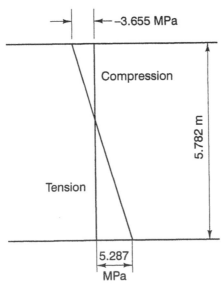

Fig. 4.9

The area of tensile steel

$$= 1.265/140 \text{ m}^2 = 9038 \text{ mm}^2$$

Use 18 bars of 25 mm diameter and 1 bar of 8 mm diameter (18 bars of 1 in diameter and 1 bar of 0.375 in diameter).

These can be arranged so that their centroid corresponds to the centroid of the triangle of tensile stress. Alternatively, while the author tends to bear this in mind to some extent, he concentrates the bars more towards the bottom, as one does in a beam. He suggests using 100 mm spacing so that the whole group covers a distance = $18 \times 0.1 = 1.8$ m (5.906 ft). These bars will have a cover of at least 25 mm (1 in) and can be placed anywhere in the depth of the plate and are usually placed alternately on top of the bottom steel for bending and beneath the top of the top steel for bending.

The minimum area of steel in 0.14 m (5.51 in) thick plate = $(0.24/100) \times 140 \times 1000 = 336$ mm²/m.

Sharing this reinforcement between the top and bottom each has $336/2 = 168$ mm²/m. Use 8 mm diameter bars at 300 mm centres (0.375 in diameter bars at 12 in centres), both top and bottom.

Maximum spacing allowed = $3 \times (140 - 25 - 4) = 333$ mm

Therefore, from the bottom, after the 1.8 m distance accommodating the tensile steel, use 8 mm diameter bars at 300 mm centres (0.375 in diameter bars at 12 in centres), top and bottom, beneath the top and above the bottom transverse bending reinforcement respectively, to act as its distribution steel. The stresses in Fig. 4.9 can be considered as due to an axial longitudinal tension T and moment M where T causes a uniform stress

$$= (3.655 + 5.287)/2 - 3.655 = 0.816 \text{ MPa}$$

M causes a bending stress = 4.471 MPa.

Fig. 4.10 shows the Fig. 4.9 stresses indicating this point. Therefore $T = 0.816 \times 0.14 \times 5.782 = 0.6605$ MN.

This has a cosine or parabolic distribution from maximum at mid-span to zero at the supports, and

$$M = 4.471 \times 0.14 \times 5.782^2/6 = 3.488 \text{ MN/m}$$

As previously

Maximum shear force = $4M/l = 4 \times 3.488/19 = 0.7343$ MN

then, taking a parabolic distribution of shear stress,

maximum shear stress = $1.5 \times (0.7343/0.14)/5.782 = 1.3607$ MPa (197.3 psi)

As T is zero at the supports, the maximum principal tensile stress at the supports is the maximum shear stress, i.e. 1.3607 MPa (197.3 psi), so diagonal tension reinforcement is required and = $1.3607 \times 0.14/140 \text{ m}^2/\text{m} = 1361$ mm²/m. Use say 16 mm diameter bars at 150 mm centres (or 5/8 in diameter bars at 6 in centres). These are at 45° to the longitudinal axis. The spacing can be increased towards the ridge and valley in accordance with the parabolic distribution of shear stress. In addition, the diagonal tension reinforcement can be reduced towards mid-span. For example, for a section 2 m from the support, the shear stress = $(2/9.5) \times 1.3607 = 0.2865$ MPa and

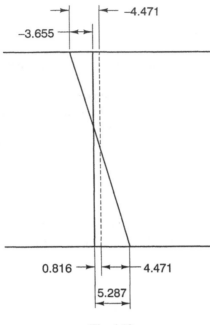

Fig. 4.10

longitudinal tensile stress = $0.816 \times [(9.5 - 2)/9.5]^2 = 0.5086$ MPa

The maximum principal tensile stress resulting from these stresses

$$= 0.5 \times (0.5086 + (0.5086^2 + 4 \times 0.2865^2)^{0.5}) = 0.6374 \text{ MPa (92.44 psi)}$$

From Section 4.3.1 one would perhaps not require diagonal tension steel at this location. Had one's design required diagonal tensile steel here, the procedure of Section 4.2.2 could be followed.

With regard to the transverse bending moments: $M1 = 0$, $M2 = -19.09$ kNm/m $= -4292$ lb ft/ft.

The loading (see Section 3.2) = 1.338 kPa (27.94 psf) of the sloping area plus self weight, which is $0.14 \times 24 = 3.360$ kPa (70.17 psf) = 4.698 kPa (98.12 psf) of the sloping area, which is 5.321 kPa (111.1 psf) of the plan area.

The maximum mid-span bending moment for free supports at the valley and ridge = $[4.698 \times 5.782 \times 10.21/2]/8 = 17.33$ kNm/m.

Figure 4.11 shows the transverse bending moment diagram. At the apex, using elastic design, BSCP114, the permissible concrete compressive stress in bending = 10 MPa (1450 psi) and the permissible tensile stress in steel = 140 MPa (20 300 psi) using the formulas from Section 4.2.2.

Using 20 mm diameter bars and 20 mm cover, the effective depth of reinforcement = $0.14 - 0.03 = 0.11$ m.

Moment of resistance = $2.14 \times 1 \times 0.11^2 = 0.02589$ MNm/m

which is greater than 19.09 kNm/m = 0.01909 MNm/m.

The area of tensile reinforcement

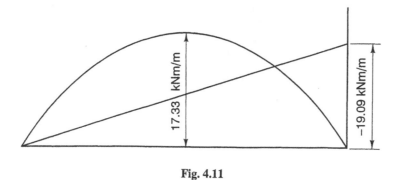

Fig. 4.11

$$= 0.01909/(0.8276 \times 0.11 \times 140) = 0.001498 \text{ m}^2/\text{m}$$
$$= 1498 \text{ mm}^2/\text{m}$$

Use 20 mm diameter bars at 200 mm centres (7/8 in diameter bars at 8 in centres).

This reinforcement requirement can be reduced towards mid-span and the maximum sagging bending moment can be calculated as in Section 4.2.2. The reinforcement system at mid-span is similar to that indicated in Fig. 4.6.

The longitudinal steel in the edge beam and the sloping plates required structurally at mid-span can be reduced towards the supports similarly to a simply supported beam carrying a uniformly distributed load. The minimum reinforcement used but not required structurally cannot be reduced.

4.3.3 Sloping plate 2–3

$$F3 = 3.666 \text{ MPa (531.7 psi)}$$

As mentioned previously, use a 1:1:2 mix of concrete as this is required for the edge beam. As $F2 = -3.655$ MPa (530.1 psi), which is less than the permissible compressive strength of 10 MPa (1450 psi), no compression steel is required.

The distribution of longitudinal stresses is shown in Fig. 4.12.

$$h = B/(2 \cos 28°) = 5.782 \text{ m (18.97 ft)}$$

The height of the neutral axis

$$= 3.666 \times 5.782/(3.655 + 3.666) = 3.666 \times 0.7898 = 2.895 \text{ m (9.498 ft)}$$

From Fig. 4.12 the longitudinal tension force can be taken as

$$= F3 \times 2.895 \times t/2 = 3.666 \times 2.895 \times 0.14/2 = 0.7429 \text{ MN}$$

The area of tensile steel $= 0.7429/140 \text{ m}^2 = 5306 \text{ mm}^2$.

Use 11 bars of 25 mm diameter (11 bars of 1 in diameter).

These can be arranged so that their centroid corresponds to the centroid of the triangle of tensile stress. While the author does bear this in mind to some extent, he alternatively concentrates the bars

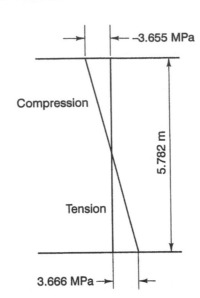

Fig. 4.12

more towards the bottom, as one does in a beam. He suggests using 100 mm spacing so that the whole group covers a distance = $10 \times 0.1 = 1.0$ m (3.381 ft). These bars will have a cover of at least 25 mm (1 in) and can be placed anywhere in the depth of the plate, and are usually placed, alternately, on top of the bottom steel for bending and beneath the top of the top steel for bending.

The minimum area of steel in 0.14 m (5.51 in) thick plate

$$= (0.24/100) \times 140 \times 1000 = 336 \text{ mm}^2/\text{m}$$

Sharing this reinforcement between the top and bottom, each has $336/2 = 168$ mm^2/m. Use 8 mm diameter bars at 300 mm centres (or 0.375 in diameter bars at 12 in centres), both top and bottom.

The maximum spacing allowed = $3 \times (140 - 25 - 4) = 333$ mm. Therefore, from the bottom, after the 1.0 m distance accommodating the tensile steel, use 8 mm diameter bars at 300 mm centres (or 0.375 in diameter bars at 12 in centres), top and bottom, beneath the top and above the bottom transverse bending reinforcement respectively to act as its distribution steel. The stresses in Fig. 4.12 can be considered as due to an axial longitudinal tension T and moment M where T causes a uniform stress

$$= (3.655 + 3.666)/2 - 3.655 = 0.0055 \text{ MPa}$$

M causes a bending stress = 3.661 MPa. Figure 4.13 shows the Fig. 4.12 stresses indicating this point. Therefore

$$T = 0.0055 \times 0.14 \times 5.782 = 0.004452 \text{ MN}$$

This has a cosine or parabolic distribution from a maximum at mid-span to zero at the supports and

$$M = 3.661 \times 0.14 \times 5.782^2/6 = 2.856 \text{ MN/m}$$

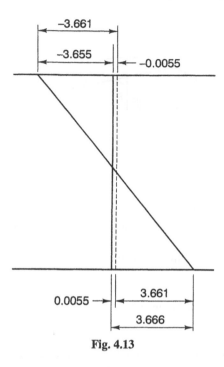

Fig. 4.13

As previously

maximum shear force = $4M/l = 4 \times 2.856/19 = 0.6013$ MN

then taking a parabolic distribution of shear stress

maximum shear stress = $1.5 \times (0.6013/0.14)/5.782 = 1.114$ MPa (161.6 psi)

As T is zero at the supports, the maximum principal tensile stress at the supports is the maximum shear stress, i.e. 1.114 MPa (161.6 psi) so diagonal tension reinforcement is required and diagonal tensile reinforcement

$= 1.114 \times 0.14/140$ m²/m $= 1114$ mm²/m

Use say 16 mm diameter bars at 175 mm centres (or 5/8 in diameter bars at 7 in centres). These are at 45° to the longitudinal axis. The spacing can be increased towards the ridge and valley in accordance with the parabolic distribution of shear stress. Also, the diagonal tension reinforcement can be reduced towards mid-span. For example for a section 2 m from the support, the maximum shear stress = $(2/9.5) \times 1.114 = 0.2345$ MPa and the longitudinal tensile stress

$= 0.0055 \times [(9.5 - 2)/9.5]^2 = 0.003428$ MPa

The maximum principal tensile stress resulting from these stresses

$= 0.5 \times (0.003428 + (0.003428^2 + 4 \times 0.2345^2)^{0.5}$
$= 0.2362$ MPa (34.26 psi)

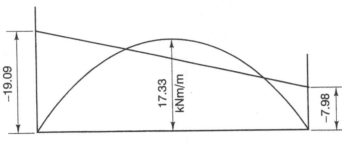

Fig. 4.14

From Section 4.3.1 one would not require diagonal tension steel at this location.

Had one's design required diagonal tensile steel here, the procedure of Section 4.2.2 could be followed.

With regard to the transverse bending moments

$$M1 = 0$$
$$M2 = -19.09 \text{ kNm/m} = -4292 \text{ lb ft/ft}$$
$$M3 = -7.98 \text{ kNm/m} = -1794 \text{ lb ft/ft}$$

The maximum mid-span bending moment for free supports at the valley and ridge = [4.698 × 5.782 × 10.21/2]/8 = 17.33 kNm/m. Figure 4.14 shows the transverse bending moment diagram. At the central valley using the elastic design, BSCP114, the permissible concrete compressive stress in bending = 10 MPa (1450 psi) and the permissible tensile stress in steel = 140 MPa (20 300 psi), and the formulas from Section 4.2.2 apply.

Using 20 mm diameter bars and 20 mm cover, the effective depth of reinforcement = 0.14 − 0.03 = 0.11 m.

$$\text{Moment of resistance} = 2.14 \times 1 \times 0.11^2 = 0.02589 \text{ MNm/m}$$

which is greater than 19.09 kNm/m = 0.01909 MNm/m.

The area of tensile reinforcement at the apex is the same as calculated in Section 4.3.2, i.e. use 20 mm diameter bars at 200 mm centres (or 7/8 in diameter bars at 8 in centres).

The area of tensile reinforcement at the valley

$$= 0.00798/(0.8276 \times 0.11 \times 140) = 0.0006261 \text{ m}^2/\text{m}$$
$$= 626.1 \text{ mm}^2/\text{m}$$

Use 16 mm diameter bars at 300 mm centres (or 5/8 in diameter bars at 12 in centres).

This reinforcement can be reduced towards mid-span and the maximum sagging bending moment can be calculated as in Section 4.2.2. The reinforcement system at mid-span is similar to that indicated in Fig. 4.6.

The longitudinal steel in the edge beam and sloping plates required structurally at mid-span can be reduced towards the supports similarly to a simply supported beam carrying a uniformly distributed load. The minimum reinforcement used but not required structurally cannot be reduced.

4.4   Type CUUEB (as shown in Figs A3 and A4; see also A2) but with many more plates (as shown in Fig. 1.1)

The data are as in Section 4.3 but with more internal plates. For approximate estimating purposes, the edge beam and the plates 1–2 and 2–3 can be designed as in Section 4.3. The extra internal plates can be reinforced as per plate 2–3, see Section 4.3.3, apart from the transverse mid-span sagging moments, because the moments at the folds may be less. The transverse mid-span sagging moment could, perhaps, for approximate estimating purposes, be taken as $17.33 - 7.98 = 9.35$ kNm/m. Looking at a scheme with eight plates, in Section 9.2.4 of Ref. 4.2, the transverse moments at the folds after fold 3 do not seem to decrease. On this limited basis, the fold moments beyond fold 3 in this example should not decrease, making the sagging moment of 9.35 kNm/m conservative. The maximum longitudinal tensile stress in a valley for the scheme of eight plates in Ref. 4.2 does increase for some internal plates but only by about 10%. The designer must make up his or her own mind whether or not he or she is happy to save time by using the tables for schemes with numerous plates.

4.5   Type PUEB (as shown in Figs A1 and A2)

The data are the same as for Section 4.2, except that the edge beams are propped similarly to that as shown in Fig. 1.10.

Thus, this folded plate lies between Tables PUEB 112 and 118. Using linear interpolation:

$$d = 1.25 + (1.28 - 1.25) \times [(19.0 - 18.69)/(19.25 - 18.69)]$$
$$= 1.25 + (1.28 - 1.25) \times [0.5536] = 1.267 \text{ m } (4.157 \text{ ft})$$
$$F0 = 0.76 + (0.79 - 0.76) \times [0.5536] = 0.7766 \text{ MPa } (112.6 \text{ psi})$$
$$F1 = F0$$
$$F2 = -\{1.29 + (1.36 - 1.29) \times [0.5536]\} = -1.329 \text{ MPa } (192.7 \text{ psi})$$
$$M1 = 0$$
$$M2 = -15.65 \text{ kNm/m}$$

4.5.1 Edge beam

Using BSCP114, for F0 and F1, a 1:1.5:3 mix of concrete, with permissible compressive stress $= 8.5$ MPa (1233 psi) (cube strength at 28 days $= 25.5$ MPa (3698 psi), about a USA cylinder strength of 22 MPa (3190 psi)) is used because this strength is required later for the transverse moments. No compression reinforcement is required. The longitudinal tension force

$$= 0.7766 \times 1.267 \times 0.23 = 0.2263 \text{ MN}$$

The permissible tensile stress in the steel $= 140$ MPa. The area of steel required $= 0.2263/140$ m^2 $= 1616$ mm^2. As this area of steel should be distributed uniformly in the edge beam, use four bars which (from Table 3.2 of Ref. 4.4) need to be of 25 mm diameter (1 in diameter). These bars can be placed one in each corner of the beam.

The minimum area of reinforcement in a 230 mm thick member (Section 3.6) $= 230 \times 1000 \times 0.24/100 = 552$ mm^2/m or, per side, $= 552/2 = 276$ mm^2/m. Thus, use 8 mm diameter bars at 175 mm centres at each side of the member. The maximum spacing of reinforcement allowed

$$= 3 \times (230 - 25 - 4) = 603 \text{ mm}$$

Nominal stirrups (links) must be used, say 10 mm diameter at 230 mm centres (or 0.375 in diameter at 9 in centres). This provides more than the minimum requirement of 552 mm²/m calculated previously.

It has been assumed that the edge beam is supported by a line reaction. If it were supported by a continuous wall, for example, it would not require further design. However, in this example, the edge beam is supported by columns.

The author once worked for a company which designed numerous shell roofs. The experience indicated that the columns should be no further apart than $2B/3$. Ideally they should be as close together as possible to please the analysis. 3.048 m (10 ft) centres were considered ideally very close and 4.572 m (15 ft) centres were happily tolerated. Ideally there should be a central column.

In the case of our present example $2B/3 = 2 \times 10.21/3 = 6.807$ m (22.33 ft). Using a central column only is therefore insufficient. Consequently, suppose three internal columns are used. This makes their spacing $= 19/4 = 4.75$ m (15.58 ft).

The beam has to carry its self weight, and past experience has ensured it is designed to carry about $B/8$ of the weight of the adjacent folded plate. As regards forces, the folded plate carries itself, but elastically some of its weight may be imposed on the edge beam between column supports.

The loading on the folded plate (see Section 3.2) = 1.338 kPa (27.94 psf) of the sloping area plus self weight, which is $0.14 \times 24 = 3.360$ kPa (70.17 psf). Therefore, the loading on the folded plate = 4.698 kPa (98.12 psf) of the sloping area, which is 5.321 kPa (111.1 psf) of the plan area.

The loading on the beam can therefore be taken as:

$$0.23 \times 1.267 \times 24 + 5.321 \times 10.21/8 = 6.994 + 6.791 = 13.78 \text{ kN/m (at a support)}$$

The beam is continuous over four spans, so using Table 6.1 of Ref. 4.4, the maximum bending moment assuming the loading mainly approximates to dead loading

$$= 0.107 \times 13.78 \times 4.75^2 = 33.27 \text{ kNm} = 0.03327 \text{ MNm (at a support)}$$

Take the effective depth of the steel $= 1.267 - 0.025 - 0.013 = 1.229$ m.

Using the data given later for a 1:1.5:3 mix: the moment of resistance $= 1.704 \times 0.23 \times 1.229^2 = 0.5920$ MNm, which is much greater than 0.03327 MNm and is satisfactory. Had the edge beam required compression reinforcement it would be designed as for any reinforced concrete beam.

The area of reinforcement required $= 0.03327/(140 \times 0.8411 \times 1.229)$ m² $= 229.9$ mm². Use one 20 mm diameter bar, which can fit between the two 25 mm diameter bars previously required to resist longitudinal tension; similarly for mid-spans.

Using Table 6.2 of Ref. 4.4 the maximum shear force

$$= 0.61 \times 13.78 \times 4.75 = 39.93 \text{ kN}$$

The corresponding maximum shear stress

$$= 39.93/(0.8411 \times 1.229 \times 0.23) \text{ kPa} = 0.1679 \text{ MPa (24.36 psi)}$$

For a 1:1.5:3 mix of concrete, BSCP114 recommends a permissible shear stress of 0.8 MPa (116 psi), but the author prefers the limits recommended in Ref. 4.1 of 0.7309 MPa (106 psi) except for rectangular beams with more than 0.17% longitudinal compression reinforcement, when the limit should be 0.5447 MPa (79 psi). Whichever is used, no shear reinforcement is required. However, had shear reinforcement been required, it would be designed as for any reinforced concrete beam.

The same struts as designed in Section 4.2.1 are required to prevent lateral instability of this edge beam.

4.5.2 Sloping plate

As mentioned previously, use a 1:1.5:3 mix of concrete as this is required later for the transverse moments. As F2 = −1.329 MPa (192.7 psi), which is less than the permissible compressive strength of 8.5 MPa (1233 psi), no compression steel is required.

The distribution of longitudinal stresses is shown in Fig. 4.15, h = 5.782 m (18.97 ft), as before. The height of the neutral axis

$$= 0.7766 \times 5.782/(0.7766 + 1.329) = 0.7766 \times 2.746 = 2.133 \text{ m } (6.997 \text{ ft})$$

From Fig. 4.15 the longitudinal tension force can be taken as

$$= F1 \times 2.133 \times t/2 = 0.7766 \times 2.133 \times 0.14/2 = 0.1160 \text{ MN}$$

The area of tensile steel

$$= 0.116/140 \text{ m}^2 = 828.2 \text{ mm}^2$$

Use eight bars of 12 mm diameter (eight bars of 1/2 in diameter). These can be arranged so that their centroid corresponds to the centroid of the triangle of tensile stress. The author, however, tends to bear this in mind to some extent, but concentrates the bars more towards the bottom, as one does in a beam. He suggests using 100 mm spacing so that the whole group covers a distance = 8 × 0.1 = 0.8 m (2.625 ft). These bars will have a cover of at least 19 mm (3/4 in) and can be placed anywhere in the depth of the plate and are usually placed alternately on top of the bottom steel for bending and beneath the top of the top steel for bending. The minimum area of steel in 0.14 m (5.51 in) thick plate

$$= (0.24/100) \times 140 \times 1000 = 336 \text{ mm}^2/\text{m}$$

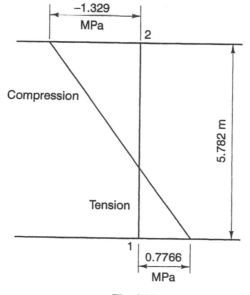

Fig. 4.15

Sharing the reinforcement between top and bottom, each has 336/2 = 168 mm²/m.

Use 8 mm diameter bars at 300 mm centres (or 0.375 in diameter bars at 12 in centres), both top and bottom.

$$\text{Maximum spacing allowed} = 3 \times (140 - 25 - 4) = 333 \text{ mm}$$

Therefore, from the bottom, after the 0.8 m distance accommodating the tensile steel, use 8 mm diameter bars at 300 mm centres top and bottom, beneath the top and above the bottom transverse bending reinforcement respectively, to act as its distribution steel.

The stresses in Fig. 4.15 can be considered as due to an axial longitudinal compression C and moment M where C causes a uniform stress

$$= (1.329 + 0.7766)/2 - 0.7766 = 1.053 - 0.7766 = 0.2762 \text{ MPa}$$

M causes a bending stress = $(1.329 + 0.7766)/2 = 1.053$ MPa. Fig. 4.16 shows the Fig. 4.15 stresses indicating this point. Therefore

$$C = 0.2762 \times 0.14 \times 5.782 = 0.2236 \text{ MN}$$

This has a cosine or parabolic distribution from maximum at mid-span to zero at the supports and

$$M = 1.053 \times 0.14 \times 5.782^2/6 = 0.8214 \text{ MN/m}$$

As previously

$$\text{maximum shear force} = 4M/l = 4 \times 0.8214/19 = 0.1729 \text{ MN}$$

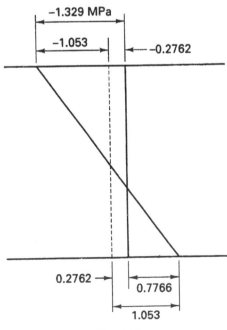

Fig. 4.16

then, taking a parabolic distribution of shear stress

maximum shear stress = $1.5 \times (0.1729/0.14)/5.782 = 0.3204$ MPa (46.47 psi)

As T is zero at the supports, the maximum principal tensile stress at the supports is the maximum shear stress, i.e. 0.3204 MPa (46.47 psi) so no diagonal tension reinforcement is required.

With regard to the transverse bending moments: $M1 = 0$ and $M2 = -15.65$ kNm/m. The loading (see Section 3.2) is 1.338 kPa (27.94 psf) of sloping area plus self weight, which is $0.14 \times 24 = 3.360$ kPa (70.17 psf). Therefore the loading is 4.698 kPa (98.12 psf) of the sloping area, which is 5.321 kPa (111.1 psf) of the plan area.

The maximum mid-span bending moment for free supports at the valley and ridge = $[4.698 \times 5.782 \times 10.21/2]/8 = 17.33$ kNm/m. Figure 4.17 shows the transverse bending moment diagram. Using, as previously, a BSCP114, 1:1.5:3 mix of concrete, permissible compressive stress = 8.5 MPa (1233 psi) (cube strength at 28 days = 25.5 MPa (3698 psi), roughly a USA cylinder strength of 22 MPa (3190 psi)), a permissible tensile stress in steel = 140 MPa (20 300 psi) and formulas from Ref. 4.4, then

$\alpha_f = 140/8.5 = 16.47$, $\alpha_e = 15$, $x_1 = 15/(16.47 + 15) = 0.4766$
$z_1 = 1 - x_1/3 = 0.8411$
$K = 0.5 \times 8.5 \times 0.4766 \times 0.8411 = 1.704$ MPa

Using 20 mm diameter bars and 20 mm cover, the effective depth of reinforcement = $0.14 - 0.03 = 0.11$ m.

Moment of resistance = $1.704 \times 1 \times 0.11^2 = 0.02062$ MNm/m

which is greater than any of the moments shown in Fig. 4.17. At the ridge, the area of tensile reinforcement

$= 0.01565/(0.8411 \times 0.11 \times 140) = 0.001208$ m²/m = 1208 mm²/m

Use 20 mm diameter bars at 250 mm centres (or 7/8 in diameter bars at 10 in centres).

This can be reduced towards mid-span and the maximum sagging bending moment can be calculated as in Section 4.2.2.

Figure 4.6 indicates (not to scale and not showing every bar) the style of the reinforcement system at mid-span.

The longitudinal steel in the edge beam and sloping plate required structurally at the mid-span can be reduced towards the supports similarly to a simply supported beam carrying a uniformly distributed load. The minimum reinforcement used but not required structurally cannot be reduced.

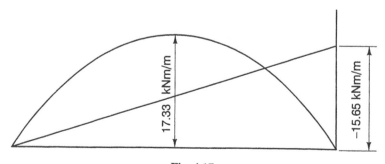

Fig. 4.17

4.6   Type UDEB (as shown in Figs A5 and A6)

Type UDEB roofs often seem to appear in USA publications. Creating the data for those who use USA/British Imperial units:

   Data: L = 60 ft (18.29 m), for loading see Section 3.2 (1.338 kPa (27.94 psf) of sloping area plus self weight).

   There are two alternatives:

4.6.1 Alternative 1

Choose B = 8.55 m and interpolate between Tables UDEB 100 and UDEB 106, or choose B = 8.81 m and interpolate between Tables UDEB 101 and UDEB 107, or choose B = 9.07 m and interpolate between Tables UDEB 102 and UDEB 108, or choose B = 9.34 m and interpolate between Tables UDEB 103 and UDEB 109, or choose B = 9.62 m and interpolate between Tables UDEB 104 and UDEB 110, or choose B = 9.91 m and interpolate between Tables UDEB 105 and UDEB 111. Then proceed as in the previous examples.

4.6.2 Alternative 2

Suppose the client demands that B = say 30 ft (9.144 m). This is within the range indicated in Alternative 1. It lies between Tables UDEB 102 (L = 18.15 and B = 9.07) and UDEB 109 (L = 18.69 and B = 9.34). The stresses and moments are close together in these two tables.

   If the interpolation uses L, then the factor to be used

$$= (18.29 - 18.15)/(18.69 - 18.15) = 0.2593$$

If interpolation uses B then the factor to be used

$$= (9.144 - 9.07)/(9.34 - 9.07) = 0.2741$$

A suggestion is to use the average = 0.2667; after all, the stresses and moments are close together in these two tables.

   Using linear interpolation:

$F0 = 5.28 + (5.66 - 5.28) \times 0.2667 = 5.381$ MPa (780.5 psi)
$F1 = - [0.15 + (0.17 - 0.15) \times 0.2667] = - 0.1553$ MPa (22.53 psi)
$F2 = - [1.46 + (1.53 - 1.46) \times 0.2667] = - 1.479$ MPa (- 214.5 psi)
$F3 = - [1.12 - (1.12 - 1.11) \times 0.2667] = - 1.117$ MPa (- 162.0 psi)
$M1 = 0$
$M2 = - [2.01 + (2.20 - 2.01) \times 0.2667] = - 2.061$ kNm/m $= - 463.4$ lb ft/ft
$M3 = - [5.37 + (5.74 - 5.37) \times 0.2667] = - 5.469$ kNm/m $= - 1229$ lb ft/ft

Then proceed as in the previous examples.

4.7   End stiffener beam

Design an end stiffener beam for the example in Section 4.2. Figure 4.18 shows a suitable beam, say 0.23 m (9.055 in) wide. This requires designing for its self weight and the parabolically distributed shear force of 0.7387 MN, which gives a maximum shear stress of 1.369 MPa. Each

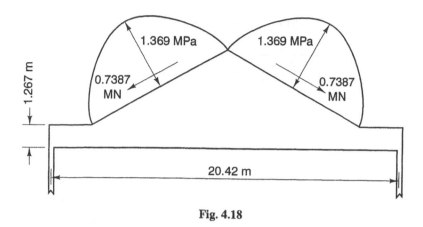

Fig. 4.18

point on the central section has a different distance to the 0.7387 MN force and therefore a different bending moment due to it. Also, this section has to resist a tension due to the horizontal component of 0.7387. Ref. 4.6 designed a beam like this.

4.8 Folded plates continuous in both directions

In Section 4.4 it was indicated how the tables might be used for plates continuous in the sideways direction. This was indicated using the type used in Section 4.3. Consider now the case of shells that are also continuous in the longitudinal direction. This will be illustrated using the same type of folded plate scheme as in Sections 4.3 and 4.4. For example, consider the scheme shown in Fig. 4.19.

The transverse bending moments might be taken as the same as those suggested in Section 4.4.

The longitudinal stresses at mid-span for a simply supported scheme, as in Section 4.4, have a parabolic or cosine distribution, as shown in Fig. 4.20. From Table 6.1 of Ref. 4.4 the distribution of the bending moment is as in Fig. 4.21. So the values of longitudinal stress that are related to a free moment and therefore 0.125 can be divided by 0.125 and multiplied by the coefficients 0.08, 0.1 and 0.025 for the various localities indicated in Fig. 4.21.

Similarly, the shear forces corresponding to Fig. 4.20 have a coefficient 0.5, so these can be divided by 0.5 and multiplied by the coefficients shown in Fig. 4.22 (which is from Table 6.2 in Ref. 4.4) for the various localities.

The designer should decide if he or she thinks this approximate method, which has been used for barrel vault roofs, is good enough. It would usually be at least good enough for estimation purposes.

4.9 Temperature movement and stresses

The coefficient of linear thermal expansion of structural concrete is approximately 0.000001 per degree Celsius (approximately 0.0000055 per degree Fahrenheit). A folded plate structure, like any other concrete structure, will alter its length and breadth according to its ambient temperature.

4.10 Shrinkage stresses

Shrinkage of concrete is a complicated problem (Ref. 4.1). The reinforcement does not shrink, consequently neither does the structure. The shrinkage stresses are relieved by a multitude of cracks, many of which cannot be seen with the naked eye (i.e. say less than 0.008 mm (0.0003

48

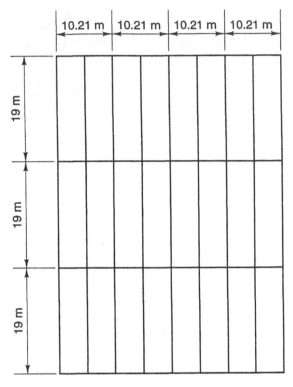

Fig. 4.19

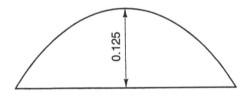

Fig. 4.20

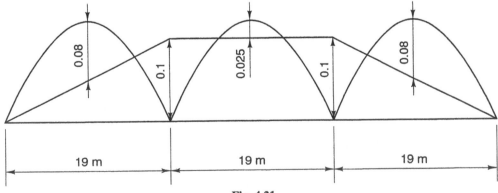

Fig. 4.21

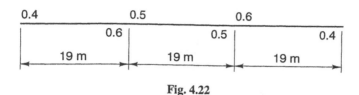

Fig. 4.22

in)). Then there is differential shrinkage due to the surface drying out faster than the inner parts of the concrete. Sunshine and/or wind on freshly placed concrete can exacerbate this effect, sometimes inciting surface cracking or crazing. Sometimes these cracks are not very deep, say 30 mm (1.2 in), but they are often visible to the naked eye and can sometimes extend to join up with cracks that are due to the reinforcement not shrinking.

With folded plates, as with most structural concrete work, shrinkage is ignored in the design, as it is considered that it is relieved by acceptable cracking.

References

4.1   Evans, R.H. and Wilby, C.B. (1963) *Concrete: Plain, Reinforced, Prestressed and Shell*, Edward Arnold, London.

4.2   Wilby, C.B. (1977) *Concrete for Structural Engineers*, Newnes-Butterworths, London.

4.3   Wilby, C.B. (1983) *Structural Concrete*, Butterworths, London, UK and Boston, USA.

4.4   Wilby, C.B. (1983) *Concrete Materials and Structures*, Cambridge University Press, Cambridge, UK, and New York, USA.

4.5   Wilby, C.B. and Khwaja, I. (1977) *Concrete Shell Roofs*, Applied Science Publishers, London.

4.6   Westbrook, R. (1984) *Structural Engineering Design in Practice*, Construction Press, London and New York.

5

Construction

5.1 Protection of concrete

It is usual in the UK to ensure water-tightness of folded plate roofs with an impermeable membrane cover for the following reasoning. There are, of course, construction joints not fitted with plastic or metal water bars, where shrinkage due to chemical action of the cement and to temperature reduction can concentrate its effect and open up cracks in these localities. Also, exposed concrete seldom dries out completely in the UK. However, about every two years there is a period of two to four weeks when there is no rain, the shade temperature is over 21° C (70° F) and there is plenty of sunshine, and exposed concrete dries out considerably. Otherwise, exposed concrete is generally damp. As the valleys act as gutters and cracks can occur here, which could be observed by photoelastic tests even if not predicted by the design, it is important that these are protected with an impermeable membrane cover.

Generally, the impermeable membrane cover would comprise three layers of built up roofing felt, with the top layer mineral finished. The latter is often dark or light green, but can be black, red or white, which is preferable for reflecting sunlight. In other words, white is better at resisting solar deterioration and heating, but it might not be favoured on the grounds of showing dirt more easily.

In some countries, some shells are just painted and presumably folded plates would be treated similarly. Furthermore, in some countries the reinforced concrete shells have no finishes whatsoever and, if a leak occurs in the very limited rainfall period, a plasterer is sent up with a bucketful of sand:cement grout and a trowel to cover over the local area of the leak.

There are some parts of the USA, for example, where frost is generally not experienced, so insulation is not vital for some buildings, and timber construction is very popular. It might be desirable in these locations to nail cedar wood tiles to strips of wood fastened to the concrete plates, and to use roofing felt or asphalt in the valleys. Similarly, burnt clay or concrete tiles (the former being favoured for lightness and possibly durability) might be used instead of wooden tiles.

5.2 Insulation

In the UK, thermal insulation can be effected by sticking 25 mm (1 in) thick corkboard on the concrete plates and beneath the roofing felt described in Section 5.1. An alternative to the corkboard is a 50 mm (2 in) or 38 mm (1.5 in) thick layer of vermiculite concrete, but this must be dried out, which can be difficult with the UK climate, before the roofing felt is laid. One technique is to waterproof the vermiculite with a thin sand and cement layer and then stick on the roofing felt when dried out, on a dry day. If it rains during the day the sand and cement water-proofing will be satisfactory until the rain ceases enough for the roofing felt contract to be continued. Another technique is to lay the roofing felt but not for about a distance of 460 mm (1.5 ft) from the valley. This valley area is then rapidly completed when the vermiculite has dried out.

Insulation can also be effected with insulating boards beneath the plates; however, insulation is better placed on top to reduce the temperature effects on the concrete plates as well as to insulate the building below.

In some countries insulation is not used.

5.3 Wall cladding

The edge beams of folded plates to brick walled buildings can be detailed as shown in Figs 5.1 to 5.3, as has similarly been done a great deal in the UK for shell roofs. The details of the tops of the beams can be used with any of the details of the bottoms of the beams. Throatings shown on the figures of this chapter can be formed, say, with 12 mm (0.5 in) half round wooden beadings.

If deflection is not significant – that is, in the cases where the edge beams are propped by intermediate columns at, say, centres (as shown in, for example, Fig. 1.10) of not more than 0.67 times the width between valleys (relating this to considerable barrel vault roof experience) – the soft material beneath the edge beam might not be used and might be replaced with slates bedded in mortar or just a layer of mortar. A desirable soft material is one that will compress and return and is resistant to water, and is available in rolls – there are such proprietary materials.

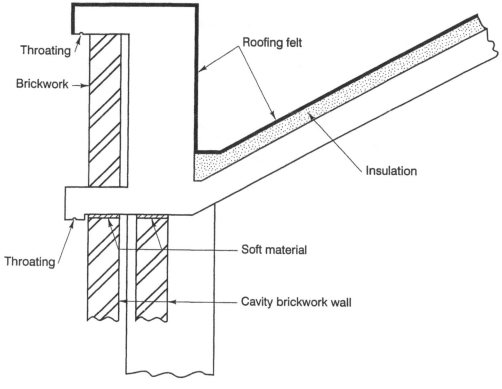

Fig. 5.1

52

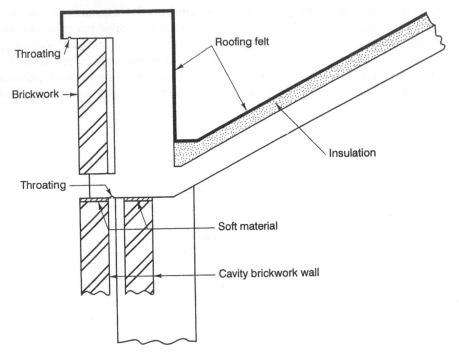

Throating

Brickwork →

Roofing felt

Insulation

Throating

Soft material

Cavity brickwork wall

Fig. 5.2

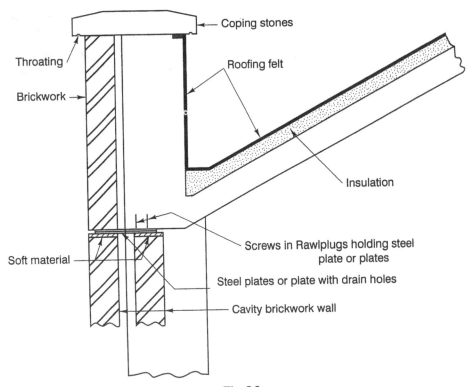

Coping stones

Throating

Brickwork →

Roofing felt

Insulation

Screws in Rawlplugs holding steel
plate or plates

Soft material —

Steel plates or plate with drain holes

Cavity brickwork wall

Fig. 5.3

If brickwork is not required to be shown on the facia of the edge beam, then the construction might be as shown in Fig. 5.4, the soft material not perhaps being used beneath the edge beam if deflection is not a worry.

If the end stiffener beam has a horizontal soffit, and brickwork is to be shown on the gable, then the brickwork outer skin can be carried on the end beam similarly to that shown in Figs 5.1 to 5.3.

For walls of 229 mm (9 in) concrete blocks or 229 mm (9 in) brickwork, or in several mainland European countries constructed of about 229 mm (9 in) by about 115 mm (4.5 in) hollow burnt clay blocks stuck together with mortar to make a wall approximately 229 mm (9 in) thick, the construction might be as shown in Fig. 5.5. The soft material does not, perhaps, have to be used beneath the edge beam if deflection is not a worry.

In the USA, timber is very popular for housing and substantial buildings (large shops and stores etc., and occasionally even bridges for carrying lorries), which would not be built of timber in the UK. In the case of California this has much to do with earthquakes. Figure 5.6 shows a possibility for using timber walls. The outer face of the wall can be of horizontal tongued and grooved boarding or cedar wood shingles. The boarding can be faced with cedar wood shingles, or a light galvanised steel wire mesh can be nailed to it before it is cement rendered. This latter is for a more robust appearance and for fire resistance. It might be painted or a coloured cement might be used.

If glass walling is used, this must slide clear of the edge beam so as not to suffer any compression; for example, see Fig. 5.7.

In the case of a cantilevered end plate the arrangement shown in Fig. 5.8 might be used.

If partition walls are built up to internal valleys, they must either be kept clear at their tops or sliding devices must be arranged so that the deflection of the valleys does not crush the partitions.

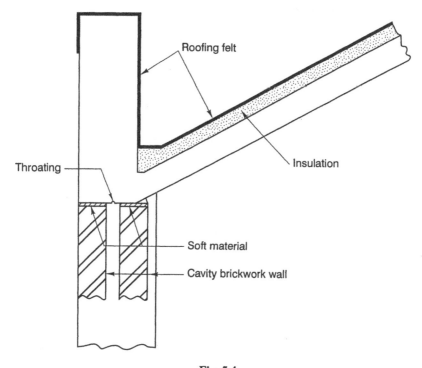

Fig. 5.4

54

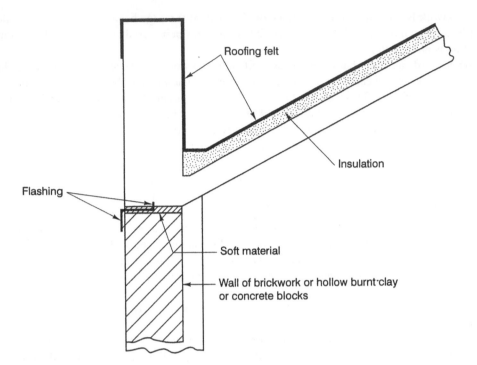

Roofing felt

Insulation

Flashing

Soft material

Wall of brickwork or hollow burnt·clay
or concrete blocks

Fig. 5.5

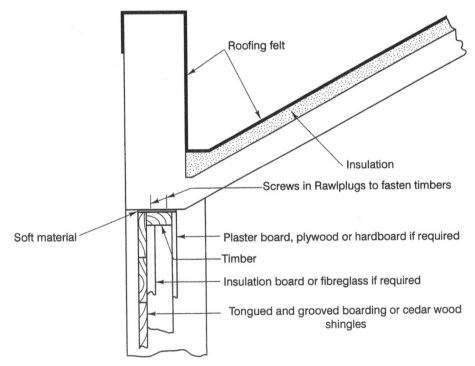

Roofing felt

Insulation

Screws in Rawlplugs to fasten timbers

Soft material

Plaster board, plywood or hardboard if required

Timber

Insulation board or fibreglass if required

Tongued and grooved boarding or cedar wood
shingles

Fig. 5.6

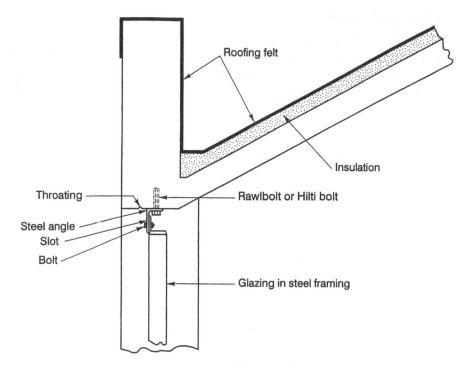

Roofing felt

Insulation

Throating

Rawlbolt or Hilti bolt

Steel angle

Slot

Bolt

Glazing in steel framing

Fig. 5.7

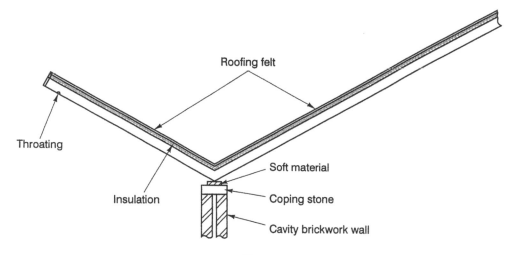

Roofing felt

Throating

Soft material

Insulation

Coping stone

Cavity brickwork wall

Fig. 5.8

5.4 Chamfers and arrises

An arris at the soffit of a valley, such as for example in Fig. 1.1, can sometimes have bits broken off it when the shutters are stripped, so a chamfer is often used here. A chamfer that is about 19 mm (0.75 in), side dimension, usually does not reduce the concrete cover to the reinforcement below the minimum requirement. Some contractors find difficulty in keeping the soffit line of the valley looking straight. This can be made easier for them by making a narrow flat portion at the soffit as shown in Fig. 5.9. This can be about 229 mm (9 in) wide without altering the analysis, assuming a similar practice as used for barrel vault roofs.

It is desirable similarly to use chamfers at the corners of columns. It is quite inexpensive and saves trouble with broken arrises on stripping shutters. Also, it makes the corners stronger, to resist damage from trucks etc. In addition, it is safer for a person bumping into the corner of the column.

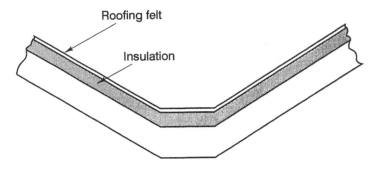

Roofing felt

Insulation

Fig. 5.9

Design Tables for Concrete Folded Plate Roofs

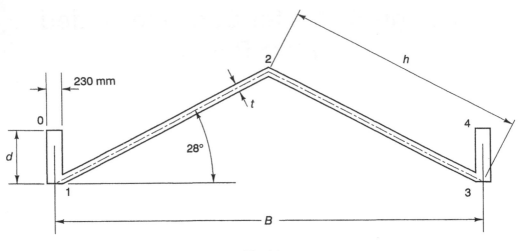

Fig. A1

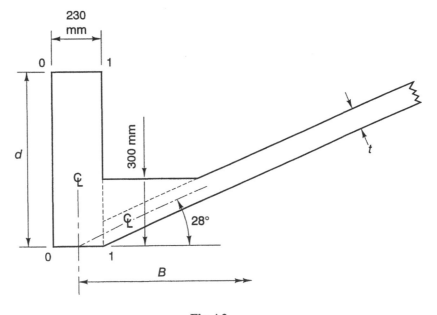

Fig. A2

Appendix 1. Design tables for folded plates types UUEB[1]

```
TABLE UUEB 1   (3 FOLDS)
L = 12.000    F0 =   -5.84
B =   5.490   F1 =    4.06
t =   0.100   F2 =   -3.01   M2 =   -7.31
d =   0.800

TABLE UUEB 2   (3 FOLDS)
L = 12.000    F0 =   -5.89
B =   5.650   F1 =    4.02
t =   0.100   F2 =   -2.95   M2 =   -7.45
d =   0.800

TABLE UUEB 3   (3 FOLDS)
L = 12.000    F0 =   -5.94
B =   5.820   F1 =    3.99
t =   0.100   F2 =   -2.90   M2 =   -7.60
d =   0.800

TABLE UUEB 4   (3 FOLDS)
L = 12.000    F0 =   -5.99
B =   5.990   F1 =    3.95
t =   0.100   F2 =   -2.85   M2 =   -7.77
d =   0.800

TABLE UUEB 5   (3 FOLDS)
L = 12.000    F0 =   -6.04
B =   6.170   F1 =    3.92
t =   0.100   F2 =   -2.80   M2 =   -7.95
d =   0.800

TABLE UUEB 6   (3 FOLDS)
L = 12.000    F0 =   -6.10
B =   6.360   F1 =    3.88
t =   0.100   F2 =   -2.75   M2 =   -8.16
d =   0.800
```

TABLE UUEB 7   (3 FOLDS)
```
L =  12.000    F0 =   -6.15
B =   6.550    F1 =    3.85
t =   0.100    F2 =   -2.70   M2 =   -8.38
d =   0.800
```

TABLE UUEB 8   (3 FOLDS)
```
L =  12.360    F0 =   -6.03
B =   5.650    F1 =    4.18
t =   0.100    F2 =   -3.09   M2 =   -7.59
d =   0.820
```

TABLE UUEB 9   (3 FOLDS)
```
L =  12.360    F0 =   -6.08
B =   5.820    F1 =    4.15
t =   0.100    F2 =   -3.04   M2 =   -7.74
d =   0.820
```

TABLE UUEB 10   (3 FOLDS)
```
L =  12.360    F0 =   -6.13
B =   5.990    F1 =    4.11
t =   0.100    F2 =   -2.98   M2 =   -7.91
d =   0.820
```

TABLE UUEB 11   (3 FOLDS)
```
L =  12.360    F0 =   -6.19
B =   6.170    F1 =    4.07
t =   0.100    F2 =   -2.93   M2 =   -8.09
d =   0.820
```

TABLE UUEB 12   (3 FOLDS)
```
L =  12.360    F0 =   -6.24
B =   6.360    F1 =    4.03
t =   0.100    F2 =   -2.88   M2 =   -8.30
d =   0.820
```

TABLE UUEB 13   (3 FOLDS)
L = 12.360   F0 =   −6.30
B =   6.550   F1 =   4.00
t =   0.100   F2 =   −2.83   M2 =   −8.51
d =   0.820

TABLE UUEB 14   (3 FOLDS)
L = 12.360   F0 =   −6.35
B =   6.750   F1 =   3.96
t =   0.100   F2 =   −2.78   M2 =   −8.76
d =   0.820

TABLE UUEB 15   (3 FOLDS)
L = 12.730   F0 =   −6.15
B =   5.820   F1 =   4.28
t =   0.100   F2 =   −3.16   M2 =   −7.78
d =   0.850

TABLE UUEB 16   (3 FOLDS)
L = 12.730   F0 =   −6.20
B =   5.990   F1 =   4.24
t =   0.100   F2 =   −3.10   M2 =   −7.95
d =   0.850

TABLE UUEB 17   (3 FOLDS)
L = 12.730   F0 =   −6.25
B =   6.170   F1 =   4.20
t =   0.100   F2 =   −3.05   M2 =   −8.13
d =   0.850

TABLE UUEB 18   (3 FOLDS)
L = 12.730   F0 =   −6.31
B =   6.360   F1 =   4.16
t =   0.100   F2 =   −2.99   M2 =   −8.34
d =   0.850

TABLE UUEB 19   (3 FOLDS)
L = 12.730   F0 =   -6.36
B =   6.550   F1 =   4.12
t =   0.100   F2 =   -2.94   M2 =   -8.56
d =   0.850

TABLE UUEB 20   (3 FOLDS)
L = 12.730   F0 =   -6.42
B =   6.750   F1 =   4.08
t =   0.100   F2 =   -2.89   M2 =   -8.81
d =   0.850

TABLE UUEB 21   (3 FOLDS)
L = 12.730   F0 =   -6.47
B =   6.950   F1 =   4.05
t =   0.100   F2 =   -2.84   M2 =   -9.07
d =   0.850

TABLE UUEB 22   (3 FOLDS)
L = 13.110   F0 =   -6.37
B =   6.000   F1 =   4.41
t =   0.100   F2 =   -3.25   M2 =   -8.11
d =   0.870

TABLE UUEB 23   (3 FOLDS)
L = 13.110   F0 =   -6.42
B =   6.180   F1 =   4.37
t =   0.100   F2 =   -3.19   M2 =   -8.29
d =   0.870

TABLE UUEB 24   (3 FOLDS)
L = 13.110   F0 =   -6.47
B =   6.370   F1 =   4.33
t =   0.100   F2 =   -3.13   M2 =   -8.50
d =   0.870

TABLE UUEB 25   (3 FOLDS)
L = 13.110   F0 =   −6.53
B =   6.560   F1 =   4.29
t =   0.100   F2 = −3.08   M2 = −8.72
d =   0.870

TABLE UUEB 26   (3 FOLDS)
L = 13.110   F0 =   −6.58
B =   6.760   F1 =   4.25
t =   0.100   F2 = −3.02   M2 = −8.96
d =   0.870

TABLE UUEB 27   (3 FOLDS)
L = 13.110   F0 =   −6.64
B =   6.960   F1 =   4.21
t =   0.100   F2 = −2.97   M2 = −9.22
d =   0.870

TABLE UUEB 28   (3 FOLDS)
L = 13.110   F0 =   −6.70
B =   7.170   F1 =   4.17
t =   0.100   F2 = −2.92   M2 = −9.51
d =   0.870

TABLE UUEB 29   (3 FOLDS)
L = 13.500   F0 =   −6.51
B =   6.180   F1 =   4.51
t =   0.100   F2 = −3.33   M2 = −8.35
d =   0.900

TABLE UUEB 30   (3 FOLDS)
L = 13.500   F0 =   −6.56
B =   6.370   F1 =   4.47
t =   0.100   F2 = −3.26   M2 = −8.56
d =   0.900

TABLE UUEB 31   (3 FOLDS)
```
L = 13.500    F0 =   -6.61
B =   6.560   F1 =    4.43
t =   0.100   F2 =   -3.21   M2 =   -8.78
d =   0.900
```

TABLE UUEB 32   (3 FOLDS)
```
L = 13.500    F0 =   -6.67
B =   6.760   F1 =    4.38
t =   0.100   F2 =   -3.15   M2 =   -9.02
d =   0.900
```

TABLE UUEB 33   (3 FOLDS)
```
L = 13.500    F0 =   -6.72
B =   6.960   F1 =    4.35
t =   0.100   F2 =   -3.10   M2 =   -9.28
d =   0.900
```

TABLE UUEB 34   (3 FOLDS)
```
L = 13.500    F0 =   -6.78
B =   7.170   F1 =    4.31
t =   0.100   F2 =   -3.04   M2 =   -9.57
d =   0.900
```

TABLE UUEB 35   (3 FOLDS)
```
L = 13.500    F0 =   -6.84
B =   7.390   F1 =    4.27
t =   0.100   F2 =   -2.99   M2 =   -9.89
d =   0.900
```

TABLE UUEB 36   (3 FOLDS)
```
L = 13.910    F0 =   -6.67
B =   6.360   F1 =    4.62
t =   0.100   F2 =   -3.41   M2 =   -8.61
d =   0.930
```

TABLE UUEB 37   (3 FOLDS)
```
L =  13.910    F0 =   -6.72
B =   6.550    F1 =    4.58
t =   0.100    F2 =   -3.35   M2 =   -8.84
d =   0.930
```

TABLE UUEB 38   (3 FOLDS)
```
L =  13.910    F0 =   -6.77
B =   6.750    F1 =    4.54
t =   0.100    F2 =   -3.29   M2 =   -9.09
d =   0.930
```

TABLE UUEB 39   (3 FOLDS)
```
L =  13.910    F0 =   -6.82
B =   6.950    F1 =    4.50
t =   0.100    F2 =   -3.23   M2 =   -9.35
d =   0.930
```

TABLE UUEB 40   (3 FOLDS)
```
L =  13.910    F0 =   -6.88
B =   7.160    F1 =    4.46
t =   0.100    F2 =   -3.18   M2 =   -9.63
d =   0.930
```

TABLE UUEB 41   (3 FOLDS)
```
L =  13.910    F0 =   -6.93
B =   7.370    F1 =    4.42
t =   0.100    F2 =   -3.13   M2 =   -9.93
d =   0.930
```

TABLE UUEB 42   (3 FOLDS)
```
L =  13.910    F0 =   -6.99
B =   7.590    F1 =    4.38
t =   0.100    F2 =   -3.07   M2 =  -10.27
d =   0.930
```

TABLE UUEB 43   (3 FOLDS)
L = 14.330   F0 =   −6.84
B =   6.560   F1 =   4.74
t =   0.100   F2 =   −3.49   M2 =   −8.93
d =   0.960

TABLE UUEB 44   (3 FOLDS)
L = 14.330   F0 =   −6.89
B =   6.760   F1 =   4.69
t =   0.100   F2 =   −3.43   M2 =   −9.18
d =   0.960

TABLE UUEB 45   (3 FOLDS)
L = 14.330   F0 =   −6.94
B =   6.960   F1 =   4.65
t =   0.100   F2 =   −3.37   M2 =   −9.44
d =   0.960

TABLE UUEB 46   (3 FOLDS)
L = 14.330   F0 =   −6.99
B =   7.170   F1 =   4.61
t =   0.100   F2 =   −3.31   M2 =   −9.73
d =   0.960

TABLE UUEB 47   (3 FOLDS)
L = 14.330   F0 =   −7.05
B =   7.390   F1 =   4.57
t =   0.100   F2 =   −3.25   M2 = −10.04
d =   0.960

TABLE UUEB 48   (3 FOLDS)
L = 14.330   F0 =   −7.11
B =   7.610   F1 =   4.52
t =   0.100   F2 =   −3.20   M2 = −10.38
d =   0.960

TABLE UUEB 49 (3 FOLDS)
L = 14.330 F0 = -7.17
B = 7.840 F1 = 4.48
t = 0.100 F2 = -3.15 M2 = -10.75
d = 0.960

TABLE UUEB 50 (3 FOLDS)
L = 14.760 F0 = -7.01
B = 6.750 F1 = 4.61
t = 0.120 F2 = -3.44 M2 = -12.42
d = 0.980

TABLE UUEB 51 (3 FOLDS)
L = 14.760 F0 = -7.08
B = 6.950 F1 = 4.57
t = 0.120 F2 = -3.37 M2 = -12.67
d = 0.980

TABLE UUEB 52 (3 FOLDS)
L = 14.760 F0 = -7.15
B = 7.160 F1 = 4.53
t = 0.120 F2 = -3.31 M2 = -12.94
d = 0.980

TABLE UUEB 53 (3 FOLDS)
L = 14.760 F0 = -7.23
B = 7.370 F1 = 4.48
t = 0.120 F2 = -3.25 M2 = -13.23
d = 0.980

TABLE UUEB 54 (3 FOLDS)
L = 14.760 F0 = -7.30
B = 7.590 F1 = 4.44
t = 0.120 F2 = -3.19 M2 = -13.55
d = 0.980

TABLE UUEB 55   (3 FOLDS)
L = 14.760   FO =   -7.38
B = 7.820   F1 =   4.40
t = 0.120   F2 =   -3.13   M2 = -13.90
d = 0.980

TABLE UUEB 56   (3 FOLDS)
L = 14.760   FO =   -7.46
B = 8.050   F1 =   4.36
t = 0.120   F2 =   -3.08   M2 = -14.28
d = 0.980

TABLE UUEB 57   (3 FOLDS)
L = 15.200   FO =   -7.20
B = 6.950   F1 =   4.74
t = 0.120   F2 =   -3.52   M2 = -12.82
d = 1.010

TABLE UUEB 58   (3 FOLDS)
L = 15.200   FO =   -7.28
B = 7.160   F1 =   4.69
t = 0.120   F2 =   -3.46   M2 = -13.10
d = 1.010

TABLE UUEB 59   (3 FOLDS)
L = 15.200   FO =   -7.35
B = 7.370   F1 =   4.65
t = 0.120   F2 =   -3.39   M2 = -13.39
d = 1.010

TABLE UUEB 60   (3 FOLDS)
L = 15.200   FO =   -7.43
B = 7.590   F1 =   4.60
t = 0.120   F2 =   -3.33   M2 = -13.71
d = 1.010

TABLE UUEB 61   (3 FOLDS)
L = 15.200   F0 =   -7.51
B =   7.820   F1 =   4.56
t =   0.120   F2 =   -3.27   M2 = -14.06
d =   1.010

TABLE UUEB 62   (3 FOLDS)
L = 15.200   F0 =   -7.58
B =   8.050   F1 =   4.52
t =   0.120   F2 =   -3.22   M2 = -14.43
d =   1.010

TABLE UUEB 63   (3 FOLDS)
L = 15.200   F0 =   -7.66
B =   8.290   F1 =   4.48
t =   0.120   F2 =   -3.16   M2 = -14.85
d =   1.010

TABLE UUEB 64   (3 FOLDS)
L = 15.660   F0 =   -7.42
B =   7.160   F1 =   4.87
t =   0.120   F2 =   -3.62   M2 = -13.27
d =   1.040

TABLE UUEB 65   (3 FOLDS)
L = 15.660   F0 =   -7.49
B =   7.370   F1 =   4.83
t =   0.120   F2 =   -3.55   M2 = -13.56
d =   1.040

TABLE UUEB 66   (3 FOLDS)
L = 15.660   F0 =   -7.57
B =   7.590   F1 =   4.78
t =   0.120   F2 =   -3.49   M2 = -13.88
d =   1.040

TABLE UUEB 67   (3 FOLDS)
L = 15.660   F0 =   -7.65
B =   7.820   F1 =   4.74
t =   0.120   F2 =   -3.42   M2 = -14.24
d =   1.040

TABLE UUEB 68   (3 FOLDS)
L = 15.660   F0 =   -7.72
B =   8.050   F1 =   4.69
t =   0.120   F2 =   -3.37   M2 = -14.61
d =   1.040

TABLE UUEB 69   (3 FOLDS)
L = 15.660   F0 =   -7.80
B =   8.290   F1 =   4.65
t =   0.120   F2 =   -3.31   M2 = -15.02
d =   1.040

TABLE UUEB 70   (3 FOLDS)
L = 15.660   F0 =   -7.89
B =   8.540   F1 =   4.60
t =   0.120   F2 =   -3.25   M2 = -15.47
d =   1.040

TABLE UUEB 71   (3 FOLDS)
L = 16.130   F0 =   -7.57
B =   7.380   F1 =   4.98
t =   0.120   F2 =   -3.70   M2 = -13.62
d =   1.080

TABLE UUEB 72   (3 FOLDS)
L = 16.130   F0 =   -7.64
B =   7.600   F1 =   4.93
t =   0.120   F2 =   -3.63   M2 = -13.94
d =   1.080

TABLE UUEB 73 (3 FOLDS)
L = 16.130   F0 =   -7.72
B = 7.830   F1 =   4.89
t = 0.120   F2 =   -3.56   M2 = -14.30
d = 1.080

TABLE UUEB 74 (3 FOLDS)
L = 16.130   F0 =   -7.79
B = 8.060   F1 =   4.84
t = 0.120   F2 =   -3.50   M2 = -14.68
d = 1.080

TABLE UUEB 75 (3 FOLDS)
L = 16.130   F0 =   -7.87
B = 8.300   F1 =   4.79
t = 0.120   F2 =   -3.44   M2 = -15.10
d = 1.080

TABLE UUEB 76 (3 FOLDS)
L = 16.130   F0 =   -7.95
B = 8.550   F1 =   4.75
t = 0.120   F2 =   -3.38   M2 = -15.56
d = 1.080

TABLE UUEB 77 (3 FOLDS)
L = 16.130   F0 =   -8.03
B = 8.810   F1 =   4.70
t = 0.120   F2 =   -3.32   M2 = -16.06
d = 1.080

TABLE UUEB 78 (3 FOLDS)
L = 16.610   F0 =   -7.80
B = 7.600   F1 =   5.12
t = 0.120   F2 =   -3.80   M2 = -14.13
d = 1.110

TABLE UUEB 79   (3 FOLDS)
L = 16.610   F0 =   -7.88
B = 7.830   F1 =   5.07
t = 0.120   F2 =   -3.73   M2 = -14.49
d = 1.110

TABLE UUEB 80   (3 FOLDS)
L = 16.610   F0 =   -7.95
B = 8.060   F1 =   5.03
t = 0.120   F2 =   -3.66   M2 = -14.87
d = 1.110

TABLE UUEB 81   (3 FOLDS)
L = 16.610   F0 =   -8.03
B = 8.300   F1 =   4.98
t = 0.120   F2 =   -3.60   M2 = -15.29
d = 1.110

TABLE UUEB 82   (3 FOLDS)
L = 16.610   F0 =   -8.11
B = 8.550   F1 =   4.93
t = 0.120   F2 =   -3.54   M2 = -15.74
d = 1.110

TABLE UUEB 83   (3 FOLDS)
L = 16.610   F0 =   -8.19
B = 8.810   F1 =   4.88
t = 0.120   F2 =   -3.47   M2 = -16.24
d = 1.110

TABLE UUEB 84   (3 FOLDS)
L = 16.610   F0 =   -8.27
B = 9.070   F1 =   4.84
t = 0.120   F2 =   -3.42   M2 = -16.77
d = 1.110

TABLE UUEB 85 (3 FOLDS)
L = 17.110   F0 =   -8.06
B = 7.830   F1 =   5.28
t = 0.120   F2 =   -3.91   M2 = -14.70
d = 1.140

TABLE UUEB 86 (3 FOLDS)
L = 17.110   F0 =   -8.13
B = 8.060   F1 =   5.23
t = 0.120   F2 =   -3.84   M2 = -15.08
d = 1.140

TABLE UUEB 87 (3 FOLDS)
L = 17.110   F0 =   -8.21
B = 8.300   F1 =   5.18
t = 0.120   F2 =   -3.77   M2 = -15.50
d = 1.140

TABLE UUEB 88 (3 FOLDS)
L = 17.110   F0 =   -8.29
B = 8.550   F1 =   5.13
t = 0.120   F2 =   -3.70   M2 = -15.95
d = 1.140

TABLE UUEB 89 (3 FOLDS)
L = 17.110   F0 =   -8.37
B = 8.810   F1 =   5.08
t = 0.120   F2 =   -3.64   M2 = -16.45
d = 1.140

TABLE UUEB 90 (3 FOLDS)
L = 17.110   F0 =   -8.45
B = 9.070   F1 =   5.03
t = 0.120   F2 =   -3.58   M2 = -16.97
d = 1.140

TABLE UUEB 91   (3 FOLDS)
L = 17.110   F0 =   -8.54
B =   9.340   F1 =   4.98
t =   0.120   F2 =   -3.52   M2 = -17.54
d =   1.140

TABLE UUEB 92   (3 FOLDS)
L = 17.620   F0 =   -8.26
B =   8.060   F1 =   5.17
t =   0.140   F2 =   -3.87   M2 = -19.59
d =   1.170

TABLE UUEB 93   (3 FOLDS)
L = 17.620   F0 =   -8.36
B =   8.300   F1 =   5.12
t =   0.140   F2 =   -3.80   M2 = -20.00
d =   1.170

TABLE UUEB 94   (3 FOLDS)
L = 17.620   F0 =   -8.46
B =   8.550   F1 =   5.07
t =   0.140   F2 =   -3.72   M2 = -20.44
d =   1.170

TABLE UUEB 95   (3 FOLDS)
L = 17.620   F0 =   -8.57
B =   8.810   F1 =   5.02
t =   0.140   F2 =   -3.65   M2 = -20.92
d =   1.170

TABLE UUEB 96   (3 FOLDS)
L = 17.620   F0 =   -8.67
B =   9.070   F1 =   4.97
t =   0.140   F2 =   -3.59   M2 = -21.43
d =   1.170

TABLE UUEB 97   (3 FOLDS)
L = 17.620   F0 =   -8.78
B = 9.340   F1 =   4.93
t = 0.140   F2 =   -3.53   M2 = -22.00
d = 1.170

TABLE UUEB 98   (3 FOLDS)
L = 17.620   F0 =   -8.89
B = 9.620   F1 =   4.88
t = 0.140   F2 =   -3.46   M2 = -22.61
d = 1.170

TABLE UUEB 99   (3 FOLDS)
L = 18.150   F0 =   -8.48
B = 8.300   F1 =   5.31
t = 0.140   F2 =   -3.97   M2 = -20.16
d = 1.210

TABLE UUEB 100   (3 FOLDS)
L = 18.150   F0 =   -8.58
B = 8.550   F1 =   5.26
t = 0.140   F2 =   -3.89   M2 = -20.61
d = 1.210

TABLE UUEB 101   (3 FOLDS)
L = 18.150   F0 =   -8.68
B = 8.810   F1 =   5.20
t = 0.140   F2 =   -3.82   M2 = -21.10
d = 1.210

TABLE UUEB 102   (3 FOLDS)
L = 18.150   F0 =   -8.78
B = 9.070   F1 =   5.15
t = 0.140   F2 =   -3.75   M2 = -21.62
d = 1.210

TABLE UUEB 103   (3 FOLDS)
L = 18.150   F0 =   -8.89
B = 9.340   F1 =   5.10
t = 0.140   F2 =   -3.68   M2 = -22.18
d = 1.210

TABLE UUEB 104   (3 FOLDS)
L = 18.150   F0 =   -8.99
B = 9.620   F1 =   5.05
t = 0.140   F2 =   -3.62   M2 = -22.80
d = 1.210

TABLE UUEB 105   (3 FOLDS)
L = 18.150   F0 =   -9.10
B = 9.910   F1 =   5.01
t = 0.140   F2 =   -3.55   M2 = -23.48
d = 1.210

TABLE UUEB 106   (3 FOLDS)
L = 18.690   F0 =   -8.70
B = 8.550   F1 =   5.45
t = 0.140   F2 =   -4.07   M2 = -20.79
d = 1.250

TABLE UUEB 107   (3 FOLDS)
L = 18.690   F0 =   -8.80
B = 8.810   F1 =   5.39
t = 0.140   F2 =   -3.99   M2 = -21.29
d = 1.250

TABLE UUEB 108   (3 FOLDS)
L = 18.690   F0 =   -8.90
B = 9.070   F1 =   5.34
t = 0.140   F2 =   -3.92   M2 = -21.81
d = 1.250

TABLE UUEB 109 (3 FOLDS)
```
L = 18.690    F0 =   -9.01
B =  9.340    F1 =    5.29
t =  0.140    F2 =   -3.85    M2 = -22.38
d =  1.250
```

TABLE UUEB 110 (3 FOLDS)
```
L = 18.690    F0 =   -9.11
B =  9.620    F1 =    5.24
t =  0.140    F2 =   -3.78    M2 = -23.00
d =  1.250
```

TABLE UUEB 111 (3 FOLDS)
```
L = 18.690    F0 =   -9.22
B =  9.910    F1 =    5.19
t =  0.140    F2 =   -3.71    M2 = -23.67
d =  1.250
```

TABLE UUEB 112 (3 FOLDS)
```
L = 18.690    F0 =   -9.33
B = 10.210    F1 =    5.14
t =  0.140    F2 =   -3.64    M2 = -24.41
d =  1.250
```

TABLE UUEB 113 (3 FOLDS)
```
L = 19.250    F0 =   -9.03
B =  8.810    F1 =    5.62
t =  0.140    F2 =   -4.19    M2 = -21.68
d =  1.280
```

TABLE UUEB 114 (3 FOLDS)
```
L = 19.250    F0 =   -9.13
B =  9.070    F1 =    5.57
t =  0.140    F2 =   -4.11    M2 = -22.20
d =  1.280
```

TABLE UUEB 115 (3 FOLDS)
```
L =  19.250    F0 =   -9.23
B =   9.340    F1 =    5.51
t =   0.140    F2 =   -4.04   M2 = -22.76
d =   1.280
```

TABLE UUEB 116 (3 FOLDS)
```
L =  19.250    F0 =   -9.34
B =   9.620    F1 =    5.46
t =   0.140    F2 =   -3.96   M2 = -23.38
d =   1.280
```

TABLE UUEB 117 (3 FOLDS)
```
L =  19.250    F0 =   -9.45
B =   9.910    F1 =    5.41
t =   0.140    F2 =   -3.89   M2 = -24.05
d =   1.280
```

TABLE UUEB 118 (3 FOLDS)
```
L =  19.250    F0 =   -9.56
B =  10.210    F1 =    5.35
t =   0.140    F2 =   -3.82   M2 = -24.78
d =   1.280
```

TABLE UUEB 119 (3 FOLDS)
```
L =  19.250    F0 =   -9.67
B =  10.520    F1 =    5.30
t =   0.140    F2 =   -3.76   M2 = -25.58
d =   1.280
```

Appendix 2. Design tables for folded plates types PUEB[1]

TABLE PUEB 1   (3 FOLDS)
L = 12.00   F0 =   0.45
B =   5.49   F1 =   0.45
t =   0.10   F2 = -0.98   M2 = -3.50
d =   0.80

TABLE PUEB 2   (3 FOLDS)
L = 12.00   F0 =   0.45
B =   5.65   F1 =   0.45
t =   0.10   F2 = -0.96   M2 = -3.71
d =   0.80

TABLE PUEB 3   (3 FOLDS)
L = 12.00   F0 =   0.44
B =   5.82   F1 =   0.44
t =   0.10   F2 = -0.94   M2 = -3.94
d =   0.80

TABLE PUEB 4   (3 FOLDS)
L = 12.00   F0 =   0.44
B =   5.99   F1 =   0.44
t =   0.10   F2 = -0.92   M2 = -4.17
d =   0.80

TABLE PUEB 5   (3 FOLDS)
L = 12.00   F0 =   0.43
B =   6.17   F1 =   0.43
t =   0.10   F2 = -0.89   M2 = -4.42
d =   0.80

TABLE PUEB 6   (3 FOLDS)
L = 12.00   F0 =   0.43
B =   6.36   F1 =   0.43
t =   0.10   F2 = -0.87   M2 = -4.70
d =   0.80

[1] See also Figs A1 and A2, p. 58

TABLE PUEB 7   (3 FOLDS)

L = 12.00	F0 = 0.43	
B = 6.55	F1 = 0.43	
t = 0.10	F2 = −0.85	M2 = −4.98
d = 0.80		

TABLE PUEB 8   (3 FOLDS)

L = 12.36	F0 = 0.47	
B = 5.65	F1 = 0.47	
t = 0.10	F2 = −1.02	M2 = −3.71
d = 0.82		

TABLE PUEB 9   (3 FOLDS)

L = 12.36	F0 = 0.46	
B = 5.82	F1 = 0.46	
t = 0.10	F2 = −0.99	M2 = −3.94
d = 0.82		

TABLE PUEB 10   (3 FOLDS)

L = 12.36	F0 = 0.46	
B = 5.99	F1 = 0.46	
t = 0.10	F2 = −0.97	M2 = −4.17
d = 0.82		

TABLE PUEB 11   (3 FOLDS)

L = 12.36	F0 = 0.45	
B = 6.17	F1 = 0.45	
t = 0.10	F2 = −0.94	M2 = −4.42
d = 0.82		

TABLE PUEB 12   (3 FOLDS)

L = 12.36	F0 = 0.45	
B = 6.36	F1 = 0.45	
t = 0.10	F2 = −0.92	M2 = −4.70
d = 0.82		

TABLE PUEB 13 **(3 FOLDS)**
L = 12.36   F0 =   0.44
B = 6.55   F1 =   0.44
t = 0.10   F2 = -0.90   M2 = -4.98
d = 0.82

TABLE PUEB 14 **(3 FOLDS)**
L = 12.36   F0 =   0.44
B = 6.75   F1 =   0.44
t = 0.10   F2 = -0.88   M2 = -5.29
d = 0.82

TABLE PUEB 15 **(3 FOLDS)**
L = 12.73   F0 =   0.48
B = 5.82   F1 =   0.48
t = 0.10   F2 = -1.04   M2 = -3.94
d = 0.85

TABLE PUEB 16 **(3 FOLDS)**
L = 12.73   F0 =   0.47
B = 5.99   F1 =   0.47
t = 0.10   F2 = -1.02   M2 = -4.17
d = 0.85

TABLE PUEB 17 **(3 FOLDS)**
L = 12.73   F0 =   0.47
B = 6.17   F1 =   0.47
t = 0.10   F2 = -0.99   M2 = -4.42
d = 0.85

TABLE PUEB 18 **(3 FOLDS)**
L = 12.73   F0 =   0.47
B = 6.36   F1 =   0.47
t = 0.10   F2 = -0.97   M2 = -4.70
d = 0.85

TABLE PUEB 19   (3 FOLDS)
L = 12.73   F0 =   0.46
B =   6.55   F1 =   0.46
t =   0.10   F2 = -0.95   M2 =   -4.98
d =   0.85

TABLE PUEB 20   (3 FOLDS)
L = 12.73   F0 =   0.46
B =   6.75   F1 =   0.46
t =   0.10   F2 = -0.92   M2 =   -5.29
d =   0.85

TABLE PUEB 21   (3 FOLDS)
L = 12.73   F0 =   0.45
B =   6.95   F1 =   0.45
t =   0.10   F2 = -0.90   M2 =   -5.61
d =   0.85

TABLE PUEB 22   (3 FOLDS)
L = 13.11   F0 =   0.49
B =   6.00   F1 =   0.49
t =   0.10   F2 = -1.08   M2 =   -4.18
d =   0.87

TABLE PUEB 23   (3 FOLDS)
L = 13.11   F0 =   0.49
B =   6.18   F1 =   0.49
t =   0.10   F2 = -1.05   M2 =   -4.44
d =   0.87

TABLE PUEB 24   (3 FOLDS)
L = 13.11   F0 =   0.49
B =   6.37   F1 =   0.49
t =   0.10   F2 = -1.02   M2 =   -4.71
d =   0.87

TABLE PUEB 25 (3 FOLDS)
L = 13.11   F0 =   0.48
B = 6.56   F1 =   0.48
t = 0.10   F2 = -1.00   M2 = -5.00
d = 0.87

TABLE PUEB 26 (3 FOLDS)
L = 13.11   F0 =   0.48
B = 6.76   F1 =   0.48
t = 0.10   F2 = -0.97   M2 = -5.31
d = 0.87

TABLE PUEB 27 (3 FOLDS)
L = 13.11   F0 =   0.47
B = 6.96   F1 =   0.47
t = 0.10   F2 = -0.95   M2 = -5.63
d = 0.87

TABLE PUEB 28 (3 FOLDS)
L = 13.11   F0 =   0.47
B = 7.17   F1 =   0.47
t = 0.10   F2 = -0.93   M2 = -5.97
d = 0.87

TABLE PUEB 29 (3 FOLDS)
L = 13.50   F0 =   0.51
B = 6.18   F1 =   0.51
t = 0.10   F2 = -1.11   M2 = -4.44
d = 0.90

TABLE PUEB 30 (3 FOLDS)
L = 13.50   F0 =   0.50
B = 6.37   F1 =   0.50
t = 0.10   F2 = -1.08   M2 = -4.71
d = 0.90

TABLE PUEB 31   (3 FOLDS)
L = 13.50   F0 =   0.50
B =   6.56   F1 =   0.50
t =   0.10   F2 = -1.05   M2 =   -5.00
d =   0.90

TABLE PUEB 32   (3 FOLDS)
L = 13.50   F0 =   0.49
B =   6.76   F1 =   0.49
t =   0.10   F2 = -1.03   M2 =   -5.31
d =   0.90

TABLE PUEB 33   (3 FOLDS)
L = 13.50   F0 =   0.49
B =   6.96   F1 =   0.49
t =   0.10   F2 = -1.00   M2 =   -5.63
d =   0.90

TABLE PUEB 34   (3 FOLDS)
L = 13.50   F0 =   0.48
B =   7.17   F1 =   0.48
t =   0.10   F2 = -0.98   M2 =   -5.97
d =   0.90

TABLE PUEB 35   (3 FOLDS)
L = 13.50   F0 =   0.48
B =   7.39   F1 =   0.48
t =   0.10   F2 = -0.95   M2 =   -6.35
d =   0.90

TABLE PUEB 36   (3 FOLDS)
L = 13.91   F0 =   0.52
B =   6.36   F1 =   0.52
t =   0.10   F2 = -1.14   M2 =   -4.70
d =   0.93

TABLE PUEB 37   (3 FOLDS)
L = 13.91   F0 =   0.52
B =   6.55   F1 =   0.52
t =   0.10   F2 =   -1.11   M2 =   -4.98
d =   0.93

TABLE PUEB 38   (3 FOLDS)
L = 13.91   F0 =   0.51
B =   6.75   F1 =   0.51
t =   0.10   F2 =   -1.09   M2 =   -5.29
d =   0.93

TABLE PUEB 39   (3 FOLDS)
L = 13.91   F0 =   0.51
B =   6.95   F1 =   0.51
t =   0.10   F2 =   -1.06   M2 =   -5.61
d =   0.93

TABLE PUEB 40   (3 FOLDS)
L = 13.91   F0 =   0.50
B =   7.16   F1 =   0.50
t =   0.10   F2 =   -1.03   M2 =   -5.96
d =   0.93

TABLE PUEB 41   (3 FOLDS)
L = 13.91   F0 =   0.50
B =   7.37   F1 =   0.50
t =   0.10   F2 =   -1.01   M2 =   -6.31
d =   0.93

TABLE PUEB 42   (3 FOLDS)
L = 13.91   F0 =   0.49
B =   7.59   F1 =   0.49
t =   0.10   F2 =   -0.98   M2 =   -6.69
d =   0.93

TABLE PUEB 43   (3 FOLDS)
```
L = 14.33    FO =    0.54
B =  6.56    F1 =    0.54
t =  0.10    F2 =   -1.17      M2 =   -5.00
d =  0.96
```

TABLE PUEB 44   (3 FOLDS)
```
L = 14.33    FO =    0.53
B =  6.76    F1 =    0.53
t =  0.10    F2 =   -1.15      M2 =   -5.31
d =  0.96
```

TABLE PUEB 45   (3 FOLDS)
```
L = 14.33    FO =    0.53
B =  6.96    F1 =    0.53
t =  0.10    F2 =   -1.12      M2 =   -5.63
d =  0.96
```

TABLE PUEB 46   (3 FOLDS)
```
L = 14.33    FO =    0.52
B =  7.17    F1 =    0.52
t =  0.10    F2 =   -1.09      M2 =   -5.97
d =  0.96
```

TABLE PUEB 47   (3 FOLDS)
```
L = 14.33    FO =    0.52
B =  7.39    F1 =    0.52
t =  0.10    F2 =   -1.06      M2 =   -6.35
d =  0.96
```

TABLE PUEB 48   (3 FOLDS)
```
L = 14.33    FO =    0.51
B =  7.61    F1 =    0.51
t =  0.10    F2 =   -1.04      M2 =   -6.73
d =  0.96
```

TABLE PUEB 49   **(3 FOLDS)**
```
L = 14.33    F0 =   0.51
B =  7.84    F1 =   0.51
t =  0.10    F2 =  -1.01    M2 =  -7.14
d =  0.96
```

TABLE PUEB 50   **(3 FOLDS)**
```
L = 14.76    F0 =   0.60
B =  6.75    F1 =   0.60
t =  0.12    F2 =  -1.19    M2 =  -6.07
d =  0.98
```

TABLE PUEB 51   **(3 FOLDS)**
```
L = 14.76    F0 =   0.60
B =  6.95    F1 =   0.60
t =  0.12    F2 =  -1.16    M2 =  -6.43
d =  0.98
```

TABLE PUEB 52   **(3 FOLDS)**
```
L = 14.76    F0 =   0.59
B =  7.16    F1 =   0.59
t =  0.12    F2 =  -1.14    M2 =  -6.83
d =  0.98
```

TABLE PUEB 53   **(3 FOLDS)**
```
L = 14.76    F0 =   0.58
B =  7.37    F1 =   0.58
t =  0.12    F2 =  -1.11    M2 =  -7.23
d =  0.98
```

TABLE PUEB 54   **(3 FOLDS)**
```
L = 14.76    F0 =   0.58
B =  7.59    F1 =   0.58
t =  0.12    F2 =  -1.08    M2 =  -7.67
d =  0.98
```

TABLE PUEB 55   (3 FOLDS)
```
L = 14.76    F0 =    0.57
B =   7.82   F1 =    0.57
t =   0.12   F2 =  -1.06      M2 =   -8.14
d =   0.98
```

TABLE PUEB 56   (3 FOLDS)
```
L = 14.76    F0 =    0.57
B =   8.05   F1 =    0.57
t =   0.12   F2 =  -1.03      M2 =   -8.63
d =   0.98
```

TABLE PUEB 57   (3 FOLDS)
```
L = 15.20    F0 =    0.62
B =   6.95   F1 =    0.62
t =   0.12   F2 =  -1.23      M2 =   -6.43
d =   1.01
```

TABLE PUEB 58   (3 FOLDS)
```
L = 15.20    F0 =    0.61
B =   7.16   F1 =    0.61
t =   0.12   F2 =  -1.20      M2 =   -6.83
d =   1.01
```

TABLE PUEB 59   (3 FOLDS)
```
L = 15.20    F0 =    0.61
B =   7.37   F1 =    0.61
t =   0.12   F2 =  -1.17      M2 =   -7.23
d =   1.01
```

TABLE PUEB 60   (3 FOLDS)
```
L = 15.20    F0 =    0.60
B =   7.59   F1 =    0.60
t =   0.12   F2 =  -1.14      M2 =   -7.67
d =   1.01
```

TABLE PUEB 61 (3 FOLDS)
```
L = 15.20    F0 =    0.59
B =  7.82    F1 =    0.59
t =  0.12    F2 =  -1.11    M2 =  -8.14
d =  1.01
```

TABLE PUEB 62 (3 FOLDS)
```
L = 15.20    F0 =    0.59
B =  8.05    F1 =    0.59
t =  0.12    F2 =  -1.09    M2 =  -8.63
d =  1.01
```

TABLE PUEB 63 (3 FOLDS)
```
L = 15.20    F0 =    0.58
B =  8.29    F1 =    0.58
t =  0.12    F2 =  -1.06    M2 =  -9.15
d =  1.01
```

TABLE PUEB 64 (3 FOLDS)
```
L = 15.66    F0 =    0.64
B =  7.16    F1 =    0.64
t =  0.12    F2 =  -1.27    M2 =  -6.83
d =  1.04
```

TABLE PUEB 65 (3 FOLDS)
```
L = 15.66    F0 =    0.63
B =  7.37    F1 =    0.63
t =  0.12    F2 =  -1.24    M2 =  -7.23
d =  1.04
```

TABLE PUEB 66 (3 FOLDS)
```
L = 15.66    F0 =    0.63
B =  7.59    F1 =    0.63
t =  0.12    F2 =  -1.21    M2 =  -7.67
d =  1.04
```

TABLE PUEB 67   (3 FOLDS)
L = 15.66   F0 =   0.62
B =   7.82   F1 =   0.62
t =   0.12   F2 = -1.18   M2 =   -8.14
d =   1.04

TABLE PUEB 68   (3 FOLDS)
L = 15.66   F0 =   0.61
B =   8.05   F1 =   0.61
t =   0.12   F2 = -1.15   M2 =   -8.63
d =   1.04

TABLE PUEB 69   (3 FOLDS)
L = 15.66   F0 =   0.61
B =   8.29   F1 =   0.61
t =   0.12   F2 = -1.12   M2 =   -9.15
d =   1.04

TABLE PUEB 70   (3 FOLDS)
L = 15.66   F0 =   0.60
B =   8.54   F1 =   0.60
t =   0.12   F2 = -1.09   M2 =   -9.71
d =   1.04

TABLE PUEB 71   (3 FOLDS)
L = 16.13   F0 =   0.65
B =   7.38   F1 =   0.65
t =   0.12   F2 = -1.30   M2 =   -7.25
d =   1.08

TABLE PUEB 72   (3 FOLDS)
L = 16.13   F0 =   0.65
B =   7.60   F1 =   0.65
t =   0.12   F2 = -1.27   M2 =   -7.69
d =   1.08

TABLE PUEB 73 **(3 FOLDS)**
L = 16.13   F0 =   0.64
B =   7.83   F1 =   0.64
t =   0.12   F2 = -1.24   M2 =   -8.17
d =   1.08

TABLE PUEB 74 **(3 FOLDS)**
L = 16.13   F0 =   0.63
B =   8.06   F1 =   0.63
t =   0.12   F2 = -1.21   M2 =   -8.65
d =   1.08

TABLE PUEB 75 **(3 FOLDS)**
L = 16.13   F0 =   0.63
B =   8.30   F1 =   0.63
t =   0.12   F2 = -1.18   M2 =   -9.17
d =   1.08

TABLE PUEB 76 **(3 FOLDS)**
L = 16.13   F0 =   0.62
B =   8.55   F1 =   0.62
t =   0.12   F2 = -1.15   M2 =   -9.74
d =   1.08

TABLE PUEB 77 **(3 FOLDS)**
L = 16.13   F0 =   0.61
B =   8.81   F1 =   0.61
t =   0.12   F2 = -1.12   M2 = -10.34
d =   1.08

TABLE PUEB 78 **(3 FOLDS)**
L = 16.61   F0 =   0.67
B =   7.60   F1 =   0.67
t =   0.12   F2 = -1.34   M2 =   -7.69
d =   1.11

TABLE PUEB 79   (3 FOLDS)
L = 16.61   F0 =   0.67
B =   7.83   F1 =   0.67
t =   0.12   F2 = -1.31   M2 =   -8.17
d =   1.11

TABLE PUEB 80   (3 FOLDS)
L = 16.61   F0 =   0.66
B =   8.06   F1 =   0.66
t =   0.12   F2 = -1.28   M2 =   -8.65
d =   1.11

TABLE PUEB 81   (3 FOLDS)
L = 16.61   F0 =   0.65
B =   8.30   F1 =   0.65
t =   0.12   F2 = -1.25   M2 =   -9.17
d =   1.11

TABLE PUEB 82   (3 FOLDS)
L = 16.61   F0 =   0.65
B =   8.55   F1 =   0.65
t =   0.12   F2 = -1.22   M2 =   -9.74
d =   1.11

TABLE PUEB 83   (3 FOLDS)
L = 16.61   F0 =   0.64
B =   8.81   F1 =   0.64
t =   0.12   F2 = -1.19   M2 = -10.34
d =   1.11

TABLE PUEB 84   (3 FOLDS)
L = 16.61   F0 =   0.63
B =   9.07   F1 =   0.63
t =   0.12   F2 = -1.16   M2 = -10.96
d =   1.11

TABLE PUEB 85 (3 FOLDS)
```
L =  17.11    F0 =     0.70
B =   7.83    F1 =     0.70
t =   0.12    F2 =    -1.38      M2 =   -8.17
d =   1.14
```

TABLE PUEB 86 (3 FOLDS)
```
L =  17.11    F0 =     0.69
B =   8.06    F1 =     0.69
t =   0.12    F2 =    -1.35      M2 =   -8.65
d =   1.14
```

TABLE PUEB 87 (3 FOLDS)
```
L =  17.11    F0 =     0.68
B =   8.30    F1 =     0.68
t =   0.12    F2 =    -1.32      M2 =   -9.17
d =   1.14
```

TABLE PUEB 88 (3 FOLDS)
```
L =  17.11    F0 =     0.67
B =   8.55    F1 =     0.67
t =   0.12    F2 =    -1.28      M2 =   -9.74
d =   1.14
```

TABLE PUEB 89 (3 FOLDS)
```
L =  17.11    F0 =     0.67
B =   8.81    F1 =     0.67
t =   0.12    F2 =    -1.25      M2 = -10.34
d =   1.14
```

TABLE PUEB 90 (3 FOLDS)
```
L =  17.11    F0 =     0.66
B =   9.07    F1 =     0.66
t =   0.12    F2 =    -1.22      M2 = -10.96
d =   1.14
```

TABLE PUEB 91   (3 FOLDS)
L = 17.11   F0 =   0.65
B =   9.34   F1 =   0.65
t =   0.12   F2 = -1.19   M2 = -11.62
d =   1.14

TABLE PUEB 92   (3 FOLDS)
L = 17.62   F0 =   0.77
B =   8.06   F1 =   0.77
t =   0.14   F2 = -1.41   M2 = -9.76
d =   1.17

TABLE PUEB 93   (3 FOLDS)
L = 17.62   F0 =   0.76
B =   8.30   F1 =   0.76
t =   0.14   F2 = -1.38   M2 = -10.35
d =   1.17

TABLE PUEB 94   (3 FOLDS)
L = 17.62   F0 =   0.75
B =   8.55   F1 =   0.75
t =   0.14   F2 = -1.35   M2 = -10.98
d =   1.17

TABLE PUEB 95   (3 FOLDS)
L = 17.62   F0 =   0.74
B =   8.81   F1 =   0.74
t =   0.14   F2 = -1.31   M2 = -11.66
d =   1.17

TABLE PUEB 96   (3 FOLDS)
L = 17.62   F0 =   0.73
B =   9.07   F1 =   0.73
t =   0.14   F2 = -1.28   M2 = -12.35
d =   1.17

TABLE PUEB 97 **(3 FOLDS)**
L = 17.62   F0 =   0.72
B =   9.34   F1 =   0.72
t =   0.14   F2 =   -1.25   M2 = -13.10
d =   1.17

TABLE PUEB 98 **(3 FOLDS)**
L = 17.62   F0 =   0.72
B =   9.62   F1 =   0.72
t =   0.14   F2 =   -1.22   M2 = -13.90
d =   1.17

TABLE PUEB 99 **(3 FOLDS)**
L = 18.15   F0 =   0.79
B =   8.30   F1 =   0.79
t =   0.14   F2 =   -1.45   M2 = -10.35
d =   1.21

TABLE PUEB 100 **(3 FOLDS)**
L = 18.15   F0 =   0.78
B =   8.55   F1 =   0.78
t =   0.14   F2 =   -1.42   M2 = -10.98
d =   1.21

TABLE PUEB 101 **(3 FOLDS)**
L = 18.15   F0 =   0.77
B =   8.81   F1 =   0.77
t =   0.14   F2 =   -1.38   M2 = -11.66
d =   1.21

TABLE PUEB 102 **(3 FOLDS)**
L = 18.15   F0 =   0.76
B =   9.07   F1 =   0.76
t =   0.14   F2 =   -1.35   M2 = -12.35
d =   1.21

TABLE PUEB 103   (3 FOLDS)
L = 18.15   F0 =   0.75
B = 9.34   F1 =   0.75
t = 0.14   F2 =   -1.32   M2 = -13.10
d = 1.21

TABLE PUEB 104   (3 FOLDS)
L = 18.15   F0 =   0.74
B = 9.62   F1 =   0.74
t = 0.14   F2 =   -1.29   M2 = -13.90
d = 1.21

TABLE PUEB 105   (3 FOLDS)
L = 18.15   F0 =   0.73
B = 9.91   F1 =   0.73
t = 0.14   F2 =   -1.26   M2 = -14.75
d = 1.21

TABLE PUEB 106   (3 FOLDS)
L = 18.69   F0 =   0.81
B = 8.55   F1 =   0.81
t = 0.14   F2 =   -1.50   M2 = -10.98
d = 1.25

TABLE PUEB 107   (3 FOLDS)
L = 18.69   F0 =   0.80
B = 8.81   F1 =   0.80
t = 0.14   F2 =   -1.46   M2 = -11.66
d = 1.25

TABLE PUEB 108   (3 FOLDS)
L = 18.69   F0 =   0.79
B = 9.07   F1 =   0.79
t = 0.14   F2 =   -1.42   M2 = -12.35
d = 1.25

TABLE PUEB 109   (3 FOLDS)
L = 18.69   F0 =   0.78
B = 9.34   F1 =   0.78
t = 0.14   F2 = -1.39   M2 = -13.10
d = 1.25

TABLE PUEB 110   (3 FOLDS)
L = 18.69   F0 =   0.77
B = 9.62   F1 =   0.77
t = 0.14   F2 = -1.36   M2 = -13.90
d = 1.25

TABLE PUEB 111   (3 FOLDS)
L = 18.69   F0 =   0.76
B = 9.91   F1 =   0.76
t = 0.14   F2 = -1.32   M2 = -14.75
d = 1.25

TABLE PUEB 112   (3 FOLDS)
L = 18.69   F0 =   0.76
B = 10.21   F1 =   0.76
t = 0.14   F2 = -1.29   M2 = -15.65
d = 1.25

TABLE PUEB 113   (3 FOLDS)
L = 19.25   F0 =   0.84
B = 8.81   F1 =   0.84
t = 0.14   F2 = -1.54   M2 = -11.66
d = 1.28

TABLE PUEB 114   (3 FOLDS)
L = 19.25   F0 =   0.83
B = 9.07   F1 =   0.83
t = 0.14   F2 = -1.51   M2 = -12.35
d = 1.28

TABLE PUEB 115   (3 FOLDS)
L = 19.25   F0 =   0.82
B = 9.34   F1 =   0.82
t = 0.14   F2 =   -1.47   M2 = -13.10
d = 1.28

TABLE PUEB 116   (3 FOLDS)
L = 19.25   F0 =   0.81
B = 9.62   F1 =   0.81
t = 0.14   F2 =   -1.43   M2 = -13.90
d = 1.28

TABLE PUEB 117   (3 FOLDS)
L = 19.25   F0 =   0.80
B = 9.91   F1 =   0.80
t = 0.14   F2 =   -1.40   M2 = -14.75
d = 1.28

TABLE PUEB 118   (3 FOLDS)
L = 19.25   F0 =   0.79
B = 10.21   F1 =   0.79
t = 0.14   F2 =   -1.36   M2 = -15.65
d = 1.28

TABLE PUEB 119   (3 FOLDS)
L = 19.25   F0 =   0.78
B = 10.52   F1 =   0.78
t = 0.14   F2 =   -1.33   M2 = -16.62
d = 1.28

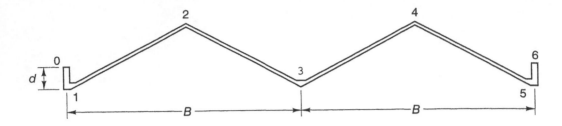

Fig. A3

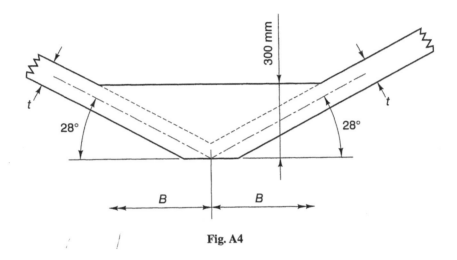

Fig. A4

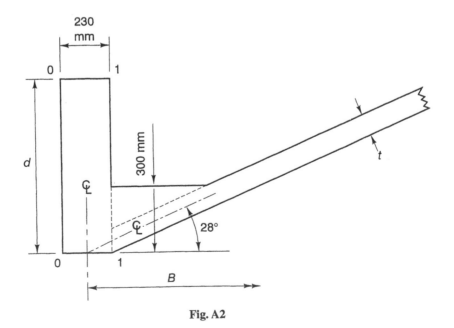

Fig. A2

Appendix 3. Design tables for folded plates types CUUEB[1]

```
TABLE CUUEB 1   (5 FOLDS)
L = 12.000    F0 =   -6.11
B =   5.490   F1 =    4.15
t =   0.100   F2 =   -3.08    M2 =   -5.35
d =   0.800   F3 =    3.17    M3 =   -1.44

TABLE CUUEB 2   (5 FOLDS)
L = 12.000    F0 =   -6.14
B =   5.650   F1 =    4.10
t =   0.100   F2 =   -3.01    M2 =   -5.50
d =   0.800   F3 =    3.09    M3 =   -1.60

TABLE CUUEB 3   (5 FOLDS)
L = 12.000    F0 =   -6.18
B =   5.820   F1 =    4.06
t =   0.100   F2 =   -2.93    M2 =   -5.66
d =   0.800   F3 =    3.00    M3 =   -1.77

TABLE CUUEB 4   (5 FOLDS)
L = 12.000    F0 =   -6.22
B =   5.990   F1 =    4.01
t =   0.100   F2 =   -2.86    M2 =   -5.82
d =   0.800   F3 =    2.91    M3 =   -1.94

TABLE CUUEB 5   (5 FOLDS)
L = 12.000    F0 =   -6.27
B =   6.170   F1 =    3.97
t =   0.100   F2 =   -2.79    M2 =   -6.01
d =   0.800   F3 =    2.83    M3 =   -2.13

TABLE CUUEB 6   (5 FOLDS)
L = 12.000    F0 =   -6.31
B =   6.360   F1 =    3.92
t =   0.100   F2 =   -2.73    M2 =   -6.21
d =   0.800   F3 =    2.75    M3 =   -2.34
```

[1] See Figs A3, A4 and A2

TABLE CUUEB 7   (5 FOLDS)

L = 12.000	F0 =	-6.36	
B = 6.550	F1 =	3.88	
t = 0.100	F2 =	-2.66	M2 = -6.43
d = 0.800	F3 =	2.67	M3 = -2.55

TABLE CUUEB 8   (5 FOLDS)

L = 12.360	F0 =	-6.30	
B = 5.650	F1 =	4.27	
t = 0.100	F2 =	-3.17	M2 = -5.58
d = 0.820	F3 =	3.26	M3 = -1.57

TABLE CUUEB 9   (5 FOLDS)

L = 12.360	F0 =	-6.34	
B = 5.820	F1 =	4.22	
t = 0.100	F2 =	-3.09	M2 = -5.74
d = 0.820	F3 =	3.17	M3 = -1.74

TABLE CUUEB 10   (5 FOLDS)

L = 12.360	F0 =	-6.38	
B = 5.990	F1 =	4.18	
t = 0.100	F2 =	-3.02	M2 = -5.90
d = 0.820	F3 =	3.08	M3 = -1.91

TABLE CUUEB 11   (5 FOLDS)

L = 12.360	F0 =	-6.42	
B = 6.170	F1 =	4.13	
t = 0.100	F2 =	-2.94	M2 = -6.09
d = 0.820	F3 =	2.99	M3 = -2.10

TABLE CUUEB 12   (5 FOLDS)

L = 12.360	F0 =	-6.47	
B = 6.360	F1 =	4.08	
t = 0.100	F2 =	-2.87	M2 = -6.29
d = 0.820	F3 =	2.91	M3 = -2.31

TABLE CUUEB 13   (5 FOLDS)
```
L =  12.360    FO =   -6.52
B =   6.550    F1 =    4.04
t =   0.100    F2 =   -2.80    M2 =   -6.51
d =   0.820    F3 =    2.83    M3 =   -2.51
```

TABLE CUUEB 14   (5 FOLDS)
```
L =  12.360    FO =   -6.56
B =   6.750    F1 =    3.99
t =   0.100    F2 =   -2.73    M2 =   -6.74
d =   0.820    F3 =    2.75    M3 =   -2.74
```

TABLE CUUEB 15   (5 FOLDS)
```
L =  12.730    FO =   -6.41
B =   5.820    F1 =    4.36
t =   0.100    F2 =   -3.24    M2 =   -5.75
d =   0.850    F3 =    3.34    M3 =   -1.73
```

TABLE CUUEB 16   (5 FOLDS)
```
L =  12.730    FO =   -6.45
B =   5.990    F1 =    4.31
t =   0.100    F2 =   -3.17    M2 =   -5.92
d =   0.850    F3 =    3.25    M3 =   -1.91
```

TABLE CUUEB 17   (5 FOLDS)
```
L =  12.730    FO =   -6.49
B =   6.170    F1 =    4.27
t =   0.100    F2 =   -3.09    M2 =   -6.11
d =   0.850    F3 =    3.16    M3 =   -2.09
```

TABLE CUUEB 18   (5 FOLDS)
```
L =  12.730    FO =   -6.54
B =   6.360    F1 =    4.22
t =   0.100    F2 =   -3.01    M2 =   -6.32
d =   0.850    F3 =    3.07    M3 =   -2.30
```

TABLE CUUEB 19   (5 FOLDS)
```
L = 12.730    F0 =  -6.58
B =  6.550    F1 =   4.17
t =  0.100    F2 =  -2.94    M2 =  -6.53
d =  0.850    F3 =   2.98    M3 =  -2.51
```

TABLE CUUEB 20   (5 FOLDS)
```
L = 12.730    F0 =  -6.63
B =  6.750    F1 =   4.12
t =  0.100    F2 =  -2.87    M2 =  -6.77
d =  0.850    F3 =   2.90    M3 =  -2.73
```

TABLE CUUEB 21   (5 FOLDS)
```
L = 12.730    F0 =  -6.68
B =  6.950    F1 =   4.08
t =  0.100    F2 =  -2.80    M2 =  -7.02
d =  0.850    F3 =   2.82    M3 =  -2.96
```

TABLE CUUEB 22   (5 FOLDS)
```
L = 13.110    F0 =  -6.63
B =  6.000    F1 =   4.49
t =  0.100    F2 =  -3.33    M2 =  -6.02
d =  0.870    F3 =   3.43    M3 =  -1.88
```

TABLE CUUEB 23   (5 FOLDS)
```
L = 13.110    F0 =  -6.67
B =  6.180    F1 =   4.44
t =  0.100    F2 =  -3.25    M2 =  -6.21
d =  0.870    F3 =   3.33    M3 =  -2.07
```

TABLE CUUEB 24   (5 FOLDS)
```
L = 13.110    F0 =  -6.72
B =  6.370    F1 =   4.39
t =  0.100    F2 =  -3.17    M2 =  -6.41
d =  0.870    F3 =   3.24    M3 =  -2.27
```

TABLE CUUEB 25   (5 FOLDS)
```
L =  13.110     F0 =   -6.76
B =   6.560     F1 =    4.34
t =   0.100     F2 =   -3.09     M2 =   -6.63
d =   0.870     F3 =    3.15     M3 =   -2.48
```

TABLE CUUEB 26   (5 FOLDS)
```
L =  13.110     F0 =   -6.81
B =   6.760     F1 =    4.29
t =   0.100     F2 =   -3.02     M2 =   -6.87
d =   0.870     F3 =    3.06     M3 =   -2.71
```

TABLE CUUEB 27   (5 FOLDS)
```
L =  13.110     F0 =   -6.86
B =   6.960     F1 =    4.25
t =   0.100     F2 =   -2.95     M2 =   -7.11
d =   0.870     F3 =    2.97     M3 =   -2.94
```

TABLE CUUEB 28   (5 FOLDS)
```
L =  13.110     F0 =   -6.91
B =   7.170     F1 =    4.20
t =   0.100     F2 =   -2.88     M2 =   -7.39
d =   0.870     F3 =    2.89     M3 =   -3.19
```

TABLE CUUEB 29   (5 FOLDS)
```
L =  13.500     F0 =   -6.77
B =   6.180     F1 =    4.59
t =   0.100     F2 =   -3.41     M2 =   -6.23
d =   0.900     F3 =    3.52     M3 =   -2.06
```

TABLE CUUEB 30   (5 FOLDS)
```
L =  13.500     F0 =   -6.81
B =   6.370     F1 =    4.54
t =   0.100     F2 =   -3.33     M2 =   -6.44
d =   0.900     F3 =    3.42     M3 =   -2.26
```

TABLE CUUEB 31   (5 FOLDS)

L =	13.500	F0 =	−6.85		
B =	6.560	F1 =	4.49		
t =	0.100	F2 =	−3.25	M2 =	−6.66
d =	0.900	F3 =	3.32	M3 =	−2.47

TABLE CUUEB 32   (5 FOLDS)

L =	13.500	F0 =	−6.90		
B =	6.760	F1 =	4.44		
t =	0.100	F2 =	−3.17	M2 =	−6.90
d =	0.900	F3 =	3.23	M3 =	−2.70

TABLE CUUEB 33   (5 FOLDS)

L =	13.500	F0 =	−6.94		
B =	6.960	F1 =	4.39		
t =	0.100	F2 =	−3.09	M2 =	−7.15
d =	0.900	F3 =	3.14	M3 =	−2.93

TABLE CUUEB 34   (5 FOLDS)

L =	13.500	F0 =	−6.99		
B =	7.170	F1 =	4.34		
t =	0.100	F2 =	−3.02	M2 =	−7.42
d =	0.900	F3 =	3.05	M3 =	−3.17

TABLE CUUEB 35   (5 FOLDS)

L =	13.500	F0 =	−7.04		
B =	7.390	F1 =	4.29		
t =	0.100	F2 =	−2.95	M2 =	−7.72
d =	0.900	F3 =	2.96	M3 =	−3.44

TABLE CUUEB 36   (5 FOLDS)

L =	13.910	F0 =	−6.92		
B =	6.360	F1 =	4.71		
t =	0.100	F2 =	−3.51	M2 =	−6.47
d =	0.930	F3 =	3.62	M3 =	−2.24

TABLE CUUEB 37   (5 FOLDS)

L = 13.910	F0 =	-6.96		
B = 6.550	F1 =	4.65		
t = 0.100	F2 =	-3.42	M2 =	-6.69
d = 0.930	F3 =	3.52	M3 =	-2.45

TABLE CUUEB 38   (5 FOLDS)

L = 13.910	F0 =	-7.01		
B = 6.750	F1 =	4.60		
t = 0.100	F2 =	-3.34	M2 =	-6.93
d = 0.930	F3 =	3.41	M3 =	-2.67

TABLE CUUEB 39   (5 FOLDS)

L = 13.910	F0 =	-7.05		
B = 6.950	F1 =	4.55		
t = 0.100	F2 =	-3.26	M2 =	-7.18
d = 0.930	F3 =	3.32	M3 =	-2.90

TABLE CUUEB 40   (5 FOLDS)

L = 13.910	F0 =	-7.10		
B = 7.160	F1 =	4.50		
t = 0.100	F2 =	-3.18	M2 =	-7.45
d = 0.930	F3 =	3.23	M3 =	-3.15

TABLE CUUEB 41   (5 FOLDS)

L = 13.910	F0 =	-7.14		
B = 7.370	F1 =	4.45		
t = 0.100	F2 =	-3.10	M2 =	-7.74
d = 0.930	F3 =	3.14	M3 =	-3.40

TABLE CUUEB 42   (5 FOLDS)

L = 13.910	F0 =	-7.19		
B = 7.590	F1 =	4.40		
t = 0.100	F2 =	-3.03	M2 =	-8.05
d = 0.930	F3 =	3.05	M3 =	-3.67

TABLE CUUEB 43   (5 FOLDS)
```
L = 14.330    F0 =  -7.09
B =  6.560    F1 =   4.82
t =  0.100    F2 = . -3.59    M2 =  -6.74
d =  0.960    F3 =   3.71     M3 =  -2.44
```

TABLE CUUEB 44   (5 FOLDS)
```
L = 14.330    F0 =  -7.13
B =  6.760    F1 =   4.77
t =  0.100    F2 =  -3.50     M2 =  -6.98
d =  0.960    F3 =   3.60     M3 =  -2.66
```

TABLE CUUEB 45   (5 FOLDS)
```
L = 14.330    F0 =  -7.17
B =  6.960    F1 =   4.72
t =  0.100    F2 =  -3.42     M2 =  -7.24
d =  0.960    F3 =   3.50     M3 =  -2.89
```

TABLE CUUEB 46   (5 FOLDS)
```
L = 14.330    F0 =  -7.22
B =  7.170    F1 =   4.66
t =  0.100    F2 =  -3.34     M2 =  -7.51
d =  0.960    F3 =   3.40     M3 =  -3.14
```

TABLE CUUEB 47   (5 FOLDS)
```
L = 14.330    F0 =  -7.27
B =  7.390    F1 =   4.61
t =  0.100    F2 =  -3.26     M2 =  -7.81
d =  0.960    F3 =   3.31     M3 =  -3.41
```

TABLE CUUEB 48   (5 FOLDS)
```
L = 14.330    F0 =  -7.32
B =  7.610    F1 =   4.56
t =  0.100    F2 =  -3.18     M2 =  -8.13
d =  0.960    F3 =   3.22     M3 =  -3.68
```

TABLE CUUEB 49 (5 FOLDS)
L = 14.330   F0 = -7.37
B = 7.840   F1 = 4.51
t = 0.100   F2 = -3.11   M2 = -8.47
d = 0.960   F3 = 3.13   M3 = -3.97

TABLE CUUEB 50 (5 FOLDS)
L = 14.760   F0 = -7.36
B = 6.750   F1 = 4.70
t = 0.120   F2 = -3.48   M2 = -9.13
d = 0.980   F3 = 3.56   M3 = -2.45

TABLE CUUEB 51 (5 FOLDS)
L = 14.760   F0 = -7.42
B = 6.950   F1 = 4.65
t = 0.120   F2 = -3.39   M2 = -9.38
d = 0.980   F3 = 3.46   M3 = -2.72

TABLE CUUEB 52 (5 FOLDS)
L = 14.760   F0 = -7.48
B = 7.160   F1 = 4.60
t = 0.120   F2 = -3.31   M2 = -9.65
d = 0.980   F3 = 3.36   M3 = -3.02

TABLE CUUEB 53 (5 FOLDS)
L = 14.760   F0 = -7.54
B = 7.370   F1 = 4.54
t = 0.120   F2 = -3.23   M2 = -9.94
d = 0.980   F3 = 3.27   M3 = -3.31

TABLE CUUEB 54 (5 FOLDS)
L = 14.760   F0 = -7.61
B = 7.590   F1 = 4.49
t = 0.120   F2 = -3.16   M2 = -10.26
d = 0.980   F3 = 3.18   M3 = -3.63

TABLE CUUEB 55   (5 FOLDS)
L = 14.760   F0 =   -7.67
B =   7.820   F1 =   4.44
t =   0.120   F2 =   -3.08   M2 = -10.60
d =   0.980   F3 =   3.10   M3 = -3.98

TABLE CUUEB 56   (5 FOLDS)
L = 14.760   F0 =   -7.74
B =   8.050   F1 =   4.39
t =   0.120   F2 =   -3.01   M2 = -10.97
d =   0.980   F3 =   3.01   M3 = -4.32

TABLE CUUEB 57   (5 FOLDS)
L = 15.200   F0 =   -7.55
B =   6.950   F1 =   4.83
t =   0.120   F2 =   -3.57   M2 = -9.47
d =   1.010   F3 =   3.66   M3 = -2.69

TABLE CUUEB 58   (5 FOLDS)
L = 15.200   F0 =   -7.61
B =   7.160   F1 =   4.77
t =   0.120   F2 =   -3.48   M2 = -9.74
d =   1.010   F3 =   3.55   M3 = -2.98

TABLE CUUEB 59   (5 FOLDS)
L = 15.200   F0 =   -7.67
B =   7.370   F1 =   4.72
t =   0.120   F2 =   -3.40   M2 = -10.03
d =   1.010   F3 =   3.46   M3 = -3.28

TABLE CUUEB 60   (5 FOLDS)
L = 15.200   F0 =   -7.74
B =   7.590   F1 =   4.66
t =   0.120   F2 =   -3.32   M2 = -10.35
d =   1.010   F3 =   3.36   M3 = -3.60

TABLE CUUEB 61   (5 FOLDS)
L = 15.200   F0 =   −7.81
B = 7.820   F1 =   4.61
t = 0.120   F2 =   −3.24   M2 = −10.69
d = 1.010   F3 =   3.27   M3 = −3.94

TABLE CUUEB 62   (5 FOLDS)
L = 15.200   F0 =   −7.87
B = 8.050   F1 =   4.56
t = 0.120   F2 =   −3.16   M2 = −11.06
d = 1.010   F3 =   3.18   M3 = −4.29

TABLE CUUEB 63   (5 FOLDS)
L = 15.200   F0 =   −7.94
B = 8.290   F1 =   4.50
t = 0.120   F2 =   −3.09   M2 = −11.45
d = 1.010   F3 =   3.09   M3 = −4.66

TABLE CUUEB 64   (5 FOLDS)
L = 15.660   F0 =   −7.77
B = 7.160   F1 =   4.96
t = 0.120   F2 =   −3.67   M2 = −9.84
d = 1.040   F3 =   3.76   M3 = −2.94

TABLE CUUEB 65   (5 FOLDS)
L = 15.660   F0 =   −7.83
B = 7.370   F1 =   4.90
t = 0.120   F2 =   −3.58   M2 = −10.13
d = 1.040   F3 =   3.65   M3 = −3.24

TABLE CUUEB 66   (5 FOLDS)
L = 15.660   F0 =   −7.89
B = 7.590   F1 =   4.85
t = 0.120   F2 =   −3.50   M2 = −10.45
d = 1.040   F3 =   3.55   M3 = −3.56

TABLE CUUEB 67　(5 FOLDS)

L = 15.660	F0 = -7.96		
B = 7.820	F1 = 4.79		
t = 0.120	F2 = -3.41	M2 = -10.80	
d = 1.040	F3 = 3.46	M3 = -3.90	

TABLE CUUEB 68　(5 FOLDS)

L = 15.660	F0 = -8.02		
B = 8.050	F1 = 4.74		
t = 0.120	F2 = -3.33	M2 = -11.16	
d = 1.040	F3 = 3.36	M3 = -4.25	

TABLE CUUEB 69　(5 FOLDS)

L = 15.660	F0 = -8.09		
B = 8.290	F1 = 4.68		
t = 0.120	F2 = -3.25	M2 = -11.56	
d = 1.040	F3 = 3.27	M3 = -4.62	

TABLE CUUEB 70　(5 FOLDS)

L = 15.660	F0 = -8.17		
B = 8.540	F1 = 4.63		
t = 0.120	F2 = -3.18	M2 = -11.99	
d = 1.040	F3 = 3.18	M3 = -5.02	

TABLE CUUEB 71　(5 FOLDS)

L = 16.130	F0 = -7.90		
B = 7.380	F1 = 5.07		
t = 0.120	F2 = -3.76	M2 = -10.16	
d = 1.080	F3 = 3.85	M3 = -3.25	

TABLE CUUEB 72　(5 FOLDS)

L = 16.130	F0 = -7.96		
B = 7.600	F1 = 5.01		
t = 0.120	F2 = -3.67	M2 = -10.49	
d = 1.080	F3 = 3.75	M3 = -3.57	

TABLE CUUEB 73 (5 FOLDS)
L = 16.130   F0 = −8.03
B = 7.830   F1 =   4.95
t = 0.120   F2 = −3.58   M2 = −10.84
d = 1.080   F3 =   3.64   M3 = −3.91

TABLE CUUEB 74 (5 FOLDS)
L = 16.130   F0 = −8.09
B = 8.060   F1 =   4.89
t = 0.120   F2 = −3.49   M2 = −11.21
d = 1.080   F3 =   3.54   M3 = −4.26

TABLE CUUEB 75 (5 FOLDS)
L = 16.130   F0 = −8.16
B = 8.300   F1 =   4.84
t = 0.120   F2 = −3.41   M2 = −11.61
d = 1.080   F3 =   3.45   M3 = −4.63

TABLE CUUEB 76 (5 FOLDS)
L = 16.130   F0 = −8.23
B = 8.550   F1 =   4.78
t = 0.120   F2 = −3.33   M2 = −12.04
d = 1.080   F3 =   3.35   M3 = −5.03

TABLE CUUEB 77 (5 FOLDS)
L = 16.130   F0 = −8.30
B = 8.810   F1 =   4.73
t = 0.120   F2 = −3.25   M2 = −12.52
d = 1.080   F3 =   3.26   M3 = −5.45

TABLE CUUEB 78 (5 FOLDS)
L = 16.610   F0 = −8.14
B = 7.600   F1 =   5.21
t = 0.120   F2 = −3.86   M2 = −10.60
d = 1.110   F3 =   3.96   M3 = −3.53

TABLE CUUEB 79   (5 FOLDS)
L = 16.610   F0 =   -8.20
B =   7.830   F1 =   5.15
t =   0.120   F2 =   -3.77   M2 = -10.95
d =   1.110   F3 =   3.85   M3 =   -3.87

TABLE CUUEB 80   (5 FOLDS)
L = 16.610   F0 =   -8.26
B =   8.060   F1 =   5.09
t =   0.120   F2 =   -3.68   M2 = -11.32
d =   1.110   F3 =   3.74   M3 =   -4.21

TABLE CUUEB 81   (5 FOLDS)
L = 16.610   F0 =   -8.33
B =   8.300   F1 =   5.03
t =   0.120   F2 =   -3.59   M2 = -11.71
d =   1.110   F3 =   3.64   M3 =   -4.59

TABLE CUUEB 82   (5 FOLDS)
L = 16.610   F0 =   -8.40
B =   8.550   F1 =   4.97
t =   0.120   F2 =   -3.51   M2 = -12.15
d =   1.110   F3 =   3.54   M3 =   -4.98

TABLE CUUEB 83   (5 FOLDS)
L = 16.610   F0 =   -8.47
B =   8.810   F1 =   4.92
t =   0.120   F2 =   -3.42   M2 = -12.62
d =   1.110   F3 =   3.44   M3 =   -5.41

TABLE CUUEB 84   (5 FOLDS)
L = 16.610   F0 =   -8.55
B =   9.070   F1 =   4.86
t =   0.120   F2 =   -3.34   M2 = -13.12
d =   1.110   F3 =   3.35   M3 =   -5.84

TABLE CUUEB 85   (5 FOLDS)
L = 17.110   F0 =   -8.39
B = 7.830   F1 =   5.36
t = 0.120   F2 =   -3.97   M2 = -11.07
d = 1.140   F3 =   4.07   M3 = -3.82

TABLE CUUEB 86   (5 FOLDS)
L = 17.110   F0 =   -8.46
B = 8.060   F1 =   5.30
t = 0.120   F2 =   -3.88   M2 = -11.44
d = 1.140   F3 =   3.96   M3 = -4.17

TABLE CUUEB 87   (5 FOLDS)
L = 17.110   F0 =   -8.52
B = 8.300   F1 =   5.24
t = 0.120   F2 =   -3.78   M2 = -11.84
d = 1.140   F3 =   3.85   M3 = -4.54

TABLE CUUEB 88   (5 FOLDS)
L = 17.110   F0 =   -8.59
B = 8.550   F1 =   5.18
t = 0.120   F2 =   -3.69   M2 = -12.27
d = 1.140   F3 =   3.74   M3 = -4.94

TABLE CUUEB 89   (5 FOLDS)
L = 17.110   F0 =   -8.66
B = 8.810   F1 =   5.12
t = 0.120   F2 =   -3.60   M2 = -12.74
d = 1.140   F3 =   3.64   M3 = -5.36

TABLE CUUEB 90   (5 FOLDS)
L = 17.110   F0 =   -8.74
B = 9.070   F1 =   5.06
t = 0.120   F2 =   -3.52   M2 = -13.23
d = 1.140   F3 =   3.54   M3 = -5.79

TABLE CUUEB 91   (5 FOLDS)
```
L =  17.110    F0 =   -8.81
B =   9.340    F1 =    5.00
t =   0.120    F2 =   -3.44    M2 = -13.77
d =   1.140    F3 =    3.45    M3 =  -6.26
```

TABLE CUUEB 92   (5 FOLDS)
```
L =  17.620    F0 =   -8.71
B =   8.060    F1 =    5.27
t =   0.140    F2 =   -3.89    M2 = -14.45
d =   1.170    F3 =    3.97    M3 =  -3.90
```

TABLE CUUEB 93   (5 FOLDS)
```
L =  17.620    F0 =   -8.79
B =   8.300    F1 =    5.21
t =   0.140    F2 =   -3.80    M2 = -14.85
d =   1.170    F3 =    3.86    M3 =  -4.33
```

TABLE CUUEB 94   (5 FOLDS)
```
L =  17.620    F0 =   -8.88
B =   8.550    F1 =    5.14
t =   0.140    F2 =   -3.71    M2 = -15.29
d =   1.170    F3 =    3.76    M3 =  -4.79
```

TABLE CUUEB 95   (5 FOLDS)
```
L =  17.620    F0 =   -8.97
B =   8.810    F1 =    5.08
t =   0.140    F2 =   -3.62    M2 = -15.76
d =   1.170    F3 =    3.65    M3 =  -5.28
```

TABLE CUUEB 96   (5 FOLDS)
```
L =  17.620    F0 =   -9.06
B =   9.070    F1 =    5.02
t =   0.140    F2 =   -3.53    M2 = -16.26
d =   1.170    F3 =    3.55    M3 =  -5.78
```

TABLE CUUEB 97   (5 FOLDS)
```
L =  17.620     F0 =    -9.16
B =   9.340     F1 =     4.97
t =   0.140     F2 =    -3.45    M2 = -16.81
d =   1.170     F3 =     3.46    M3 =  -6.32
```

TABLE CUUEB 98   (5 FOLDS)
```
L =  17.620     F0 =    -9.25
B =   9.620     F1 =     4.91
t =   0.140     F2 =    -3.37    M2 = -17.40
d =   1.170     F3 =     3.37    M3 =  -6.88
```

TABLE CUUEB 99   (5 FOLDS)
```
L =  18.150     F0 =    -8.92
B =   8.300     F1 =     5.40
t =   0.140     F2 =    -4.00    M2 = -14.95
d =   1.210     F3 =     4.08    M3 =  -4.30
```

TABLE CUUEB 100   (5 FOLDS)
```
L =  18.150     F0 =    -9.00
B =   8.550     F1 =     5.34
t =   0.140     F2 =    -3.90    M2 = -15.39
d =   1.210     F3 =     3.97    M3 =  -4.76
```

TABLE CUUEB 101   (5 FOLDS)
```
L =  18.150     F0 =    -9.09
B =   8.810     F1 =     5.27
t =   0.140     F2 =    -3.81    M2 = -15.87
d =   1.210     F3 =     3.86    M3 =  -5.25
```

TABLE CUUEB 102   (5 FOLDS)
```
L =  18.150     F0 =    -9.18
B =   9.070     F1 =     5.21
t =   0.140     F2 =    -3.72    M2 = -16.37
d =   1.210     F3 =     3.75    M3 =  -5.75
```

TABLE CUUEB 103   (5 FOLDS)
L = 18.150   F0 =   -9.27
B = 9.340   F1 =   5.15
t = 0.140   F2 =   -3.63   M2 = -16.92
d = 1.210   F3 =   3.65   M3 = -6.28

TABLE CUUEB 104   (5 FOLDS)
L = 18.150   F0 =   -9.37
B = 9.620   F1 =   5.09
t = 0.140   F2 =   -3.54   M2 = -17.51
d = 1.210   F3 =   3.55   M3 = -6.84

TABLE CUUEB 105   (5 FOLDS)
L = 18.150   F0 =   -9.46
B = 9.910   F1 =   5.03
t = 0.140   F2 =   -3.46   M2 = -18.16
d = 1.210   F3 =   3.46   M3 = -7.44

TABLE CUUEB 106   (5 FOLDS)
L = 18.690   F0 =   -9.14
B = 8.550   F1 =   5.54
t = 0.140   F2 =   -4.10   M2 = -15.49
d = 1.250   F3 =   4.19   M3 = -4.72

TABLE CUUEB 107   (5 FOLDS)
L = 18.690   F0 =   -9.22
B = 8.810   F1 =   5.47
t = 0.140   F2 =   -4.00   M2 = -15.97
d = 1.250   F3 =   4.07   M3 = -5.21

TABLE CUUEB 108   (5 FOLDS)
L = 18.690   F0 =   -9.31
B = 9.070   F1 =   5.41
t = 0.140   F2 =   -3.91   M2 = -16.48
d = 1.250   F3 =   3.96   M3 = -5.71

TABLE CUUEB 109   (5 FOLDS)
```
L =  18.690     F0 =   -9.40
B =   9.340     F1 =    5.35
t =   0.140     F2 =   -3.82     M2 = -17.03
d =   1.250     F3 =    3.86     M3 =  -6.24
```

TABLE CUUEB 110   (5 FOLDS)
```
L =  18.690     F0 =   -9.49
B =   9.620     F1 =    5.28
t =   0.140     F2 =   -3.73     M2 = -17.62
d =   1.250     F3 =    3.75     M3 =  -6.80
```

TABLE CUUEB 111   (5 FOLDS)
```
L =  18.690     F0 =   -9.59
B =   9.910     F1 =    5.22
t =   0.140     F2 =   -3.64     M2 = -18.27
d =   1.250     F3 =    3.65     M3 =  -7.40
```

TABLE CUUEB 112   (5 FOLDS)
```
L =  18.690     F0 =   -9.69
B =  10.210     F1 =    5.16
t =   0.140     F2 =   -3.55     M2 = -18.97
d =   1.250     F3 =    3.55     M3 =  -8.03
```

TABLE CUUEB 113   (5 FOLDS)
```
L =  19.250     F0 =   -9.46
B =   8.810     F1 =    5.71
t =   0.140     F2 =   -4.22     M2 = -16.21
d =   1.280     F3 =    4.31     M3 =  -5.11
```

TABLE CUUEB 114   (5 FOLDS)
```
L =  19.250     F0 =   -9.55
B =   9.070     F1 =    5.65
t =   0.140     F2 =   -4.12     M2 = -16.71
d =   1.280     F3 =    4.19     M3 =  -5.61
```

TABLE CUUEB 115   (5 FOLDS)
L = 19.250   F0 =   -9.64
B = 9.340   F1 =   5.58
t = 0.140   F2 =   -4.02   M2 = -17.26
d = 1.280   F3 =   4.08   M3 = -6.14

TABLE CUUEB 116   (5 FOLDS)
L = 19.250   F0 =   -9.73
B = 9.620   F1 =   5.52
t = 0.140   F2 =   -3.93   M2 = -17.85
d = 1.280   F3 =   3.97   M3 = -6.71

TABLE CUUEB 117   (5 FOLDS)
L = 19.250   F0 =   -9.83
B = 9.910   F1 =   5.45
t = 0.140   F2 =   -3.83   M2 = -18.49
d = 1.280   F3 =   3.86   M3 = -7.31

TABLE CUUEB 118   (5 FOLDS)
L = 19.250   F0 =   -9.93
B = 10.210   F1 =   5.39
t = 0.140   F2 =   -3.74   M2 = -19.19
d = 1.280   F3 =   3.76   M3 = -7.94

TABLE CUUEB 119   (5 FOLDS)
L = 19.250   F0 = -10.04
B = 10.520   F1 =   5.32
t = 0.140   F2 =   -3.66   M2 = -19.94
d = 1.280   F3 =   3.65   M3 = -8.61

Appendix 4. Design tables for folded plates types CPUEB[1]

TABLE CPUEB 1   (5 FOLDS)
L = 12.00   F0 =   0.28
B =   5.49   F1 =   0.69
t =   0.10   F2 = -1.67   M2 =   -3.90
d =   0.80   F3 =   2.08   M3 =   -1.33

TABLE CPUEB 2   (5 FOLDS)
L = 12.00   F0 =   0.25
B =   5.65   F1 =   0.69
t =   0.10   F2 = -1.63   M2 =   -4.07
d =   0.80   F3 =   2.03   M3 =   -1.47

TABLE CPUEB 3   (5 FOLDS)
L = 12.00   F0 =   0.23
B =   5.82   F1 =   0.69
t =   0.10   F2 = -1.59   M2 =   -4.25
d =   0.80   F3 =   1.97   M3 =   -1.63

TABLE CPUEB 4   (5 FOLDS)
L = 12.00   F0 =   0.21
B =   5.99   F1 =   0.69
t =   0.10   F2 = -1.55   M2 =   -4.44
d =   0.80   F3 =   1.92   M3 =   -1.79

TABLE CPUEB 5   (5 FOLDS)
L = 12.00   F0 =   0.18
B =   6.17   F1 =   0.69
t =   0.10   F2 = -1.51   M2 =   -4.65
d =   0.80   F3 =   1.87   M3 =   -1.96

TABLE CPUEB 6   (5 FOLDS)
L = 12.00   F0 =   0.16
B =   6.36   F1 =   0.69
t =   0.10   F2 = -1.47   M2 =   -4.87
d =   0.80   F3 =   1.81   M3 =   -2.14

[1] See also Figs A3, A4 and A2, p. 100

TABLE CPUEB 7   (5 FOLDS)
```
L = 12.00     F0 =     0.13
B =   6.55    F1 =     0.69
t =   0.10    F2 =    -1.43     M2 =    -5.10
d =   0.80    F3 =     1.76     M3 =    -2.32
```

TABLE CPUEB 8   (5 FOLDS)
```
L = 12.36     F0 =     0.29
B =   5.65    F1 =     0.71
t =   0.10    F2 =    -1.72     M2 =    -4.10
d =   0.82    F3 =     2.14     M3 =    -1.43
```

TABLE CPUEB 9   (5 FOLDS)
```
L = 12.36     F0 =     0.26
B =   5.82    F1 =     0.71
t =   0.10    F2 =    -1.68     M2 =    -4.29
d =   0.82    F3 =     2.09     M3 =    -1.59
```

TABLE CPUEB 10   (5 FOLDS)
```
L = 12.36     F0 =     0.24
B =   5.99    F1 =     0.71
t =   0.10    F2 =    -1.63     M2 =    -4.48
d =   0.82    F3 =     2.03     M3 =    -1.75
```

TABLE CPUEB 11   (5 FOLDS)
```
L = 12.36     F0 =     0.21
B =   6.17    F1 =     0.71
t =   0.10    F2 =    -1.59     M2 =    -4.69
d =   0.82    F3 =     1.98     M3 =    -1.92
```

TABLE CPUEB 12   (5 FOLDS)
```
L = 12.36     F0 =     0.19
B =   6.36    F1 =     0.71
t =   0.10    F2 =    -1.55     M2 =    -4.91
d =   0.82    F3 =     1.92     M3 =    -2.10
```

TABLE CPUEB 13   (5 FOLDS)
L = 12.36   F0 =   0.16
B =   6.55   F1 =   0.71
t =   0.10   F2 =   -1.51   M2 =   -5.14
d =   0.82   F3 =   1.87   M3 =   -2.29

TABLE CPUEB 14   (5 FOLDS)
L = 12.36   F0 =   0.14
B =   6.75   F1 =   0.71
t =   0.10   F2 =   -1.47   M2 =   -5.39
d =   0.82   F3 =   1.82   M3 =   -2.48

TABLE CPUEB 15   (5 FOLDS)
L = 12.73   F0 =   0.29
B =   5.82   F1 =   0.73
t =   0.10   F2 =   -1.77   M2 =   -4.32
d =   0.85   F3 =   2.21   M3 =   -1.55

TABLE CPUEB 16   (5 FOLDS)
L = 12.73   F0 =   0.27
B =   5.99   F1 =   0.73
t =   0.10   F2 =   -1.73   M2 =   -4.52
d =   0.85   F3 =   2.15   M3 =   -1.71

TABLE CPUEB 17   (5 FOLDS)
L = 12.73   F0 =   0.24
B =   6.17   F1 =   0.73
t =   0.10   F2 =   -1.68   M2 =   -4.73
d =   0.85   F3 =   2.09   M3 =   -1.88

TABLE CPUEB 18   (5 FOLDS)
L = 12.73   F0 =   0.22
B =   6.36   F1 =   0.73
t =   0.10   F2 =   -1.64   M2 =   -4.95
d =   0.85   F3 =   2.03   M3 =   -2.06

TABLE CPUEB 19   (5 FOLDS)
L = 12.73   F0 =   0.19
B =   6.55   F1 =   0.73
t =   0.10   F2 =   -1.60   M2 =   -5.18
d =   0.85   F3 =   1.98   M3 =   -2.25

TABLE CPUEB 20   (5 FOLDS)
L = 12.73   F0 =   0.17
B =   6.75   F1 =   0.73
t =   0.10   F2 =   -1.56   M2 =   -5.43
d =   0.85   F3 =   1.92   M3 =   -2.45

TABLE CPUEB 21   (5 FOLDS)
L = 12.73   F0 =   0.14
B =   6.95   F1 =   0.73
t =   0.10   F2 =   -1.52   M2 =   -5.69
d =   0.85   F3 =   1.87   M3 =   -2.65

TABLE CPUEB 22   (5 FOLDS)
L = 13.11   F0 =   0.30
B =   6.00   F1 =   0.75
t =   0.10   F2 =   -1.82   M2 =   -4.56
d =   0.87   F3 =   2.27   M3 =   -1.67

TABLE CPUEB 23   (5 FOLDS)
L = 13.11   F0 =   0.28
B =   6.18   F1 =   0.75
t =   0.10   F2 =   -1.78   M2 =   -4.77
d =   0.87   F3 =   2.21   M3 =   -1.85

TABLE CPUEB 24   (5 FOLDS)
L = 13.11   F0 =   0.25
B =   6.37   F1 =   0.75
t =   0.10   F2 =   -1.73   M2 =   -5.00
d =   0.87   F3 =   2.15   M3 =   -2.03

TABLE CPUEB 25   (5 FOLDS)
L = 13.11   F0 =   0.22
B =   6.56   F1 =   0.75
t =   0.10   F2 =   -1.69   M2 =   -5.23
d =   0.87   F3 =   2.09   M3 =   -2.22

TABLE CPUEB 26   (5 FOLDS)
L = 13.11   F0 =   0.20
B =   6.76   F1 =   0.75
t =   0.10   F2 =   -1.64   M2 =   -5.48
d =   0.87   F3 =   2.03   M3 =   -2.42

TABLE CPUEB 27   (5 FOLDS)
L = 13.11   F0 =   0.17
B =   6.96   F1 =   0.75
t =   0.10   F2 =   -1.60   M2 =   -5.74
d =   0.87   F3 =   1.98   M3 =   -2.63

TABLE CPUEB 28   (5 FOLDS)
L = 13.11   F0 =   0.15
B =   7.17   F1 =   0.75
t =   0.10   F2 =   -1.56   M2 =   -6.02
d =   0.87   F3 =   1.93   M3 =   -2.85

TABLE CPUEB 29   (5 FOLDS)
L = 13.50   F0 =   0.31
B =   6.18   F1 =   0.77
t =   0.10   F2 =   -1.88   M2 =   -4.81
d =   0.90   F3 =   2.34   M3 =   -1.80

TABLE CPUEB 30   (5 FOLDS)
L = 13.50   F0 =   0.28
B =   6.37   F1 =   0.77
t =   0.10   F2 =   -1.83   M2 =   -5.04
d =   0.90   F3 =   2.28   M3 =   -1.99

TABLE CPUEB 31   (5 FOLDS)
```
L =  13.50    F0 =      0.26
B =   6.56    F1 =      0.77
t =   0.10    F2 =     -1.78    M2 =   -5.27
d =   0.90    F3 =      2.21    M3 =   -2.18
```

TABLE CPUEB 32   (5 FOLDS)
```
L =  13.50    F0 =      0.23
B =   6.76    F1 =      0.77
t =   0.10    F2 =     -1.74    M2 =   -5.53
d =   0.90    F3 =      2.15    M3 =   -2.38
```

TABLE CPUEB 33   (5 FOLDS)
```
L =  13.50    F0 =      0.20
B =   6.96    F1 =      0.78
t =   0.10    F2 =     -1.69    M2 =   -5.79
d =   0.90    F3 =      2.10    M3 =   -2.59
```

TABLE CPUEB 34   (5 FOLDS)
```
L =  13.50    F0 =      0.17
B =   7.17    F1 =      0.78
t =   0.10    F2 =     -1.65    M2 =   -6.06
d =   0.90    F3 =      2.04    M3 =   -2.81
```

TABLE CPUEB 35   (5 FOLDS)
```
L =  13.50    F0 =      0.15
B =   7.39    F1 =      0.78
t =   0.10    F2 =     -1.61    M2 =   -6.36
d =   0.90    F3 =      1.98    M3 =   -3.04
```

TABLE CPUEB 36   (5 FOLDS)
```
L =  13.91    F0 =      0.32
B =   6.36    F1 =      0.79
t =   0.10    F2 =     -1.93    M2 =   -5.07
d =   0.93    F3 =      2.42    M3 =   -1.94
```

TABLE CPUEB 37   (5 FOLDS)
L = 13.91   F0 =   0.29
B = 6.55   F1 =   0.79
t = 0.10   F2 = -1.89   M2 = -5.30
d = 0.93   F3 =   2.35   M3 = -2.13

TABLE CPUEB 38   (5 FOLDS)
L = 13.91   F0 =   0.26
B = 6.75   F1 =   0.80
t = 0.10   F2 = -1.84   M2 = -5.56
d = 0.93   F3 =   2.28   M3 = -2.33

TABLE CPUEB 39   (5 FOLDS)
L = 13.91   F0 =   0.24
B = 6.95   F1 =   0.80
t = 0.10   F2 = -1.79   M2 = -5.82
d = 0.93   F3 =   2.22   M3 = -2.54

TABLE CPUEB 40   (5 FOLDS)
L = 13.91   F0 =   0.21
B = 7.16   F1 =   0.80
t = 0.10   F2 = -1.75   M2 = -6.10
d = 0.93   F3 =   2.16   M3 = -2.76

TABLE CPUEB 41   (5 FOLDS)
L = 13.91   F0 =   0.18
B = 7.37   F1 =   0.80
t = 0.10   F2 = -1.70   M2 = -6.38
d = 0.93   F3 =   2.10   M3 = -2.98

TABLE CPUEB 42   (5 FOLDS)
L = 13.91   F0 =   0.15
B = 7.59   F1 =   0.80
t = 0.10   F2 = -1.66   M2 = -6.69
d = 0.93   F3 =   2.05   M3 = -3.22

TABLE CPUEB 43   (5 FOLDS)

```
L = 14.33    F0 =     0.32
B =   6.56   F1 =     0.82
t =   0.10   F2 =    -1.99    M2 =    -5.35
d =   0.96   F3 =     2.49    M3 =    -2.09
```

TABLE CPUEB 44   (5 FOLDS)

```
L = 14.33    F0 =     0.30
B =   6.76   F1 =     0.82
t =   0.10   F2 =    -1.94    M2 =    -5.61
d =   0.96   F3 =     2.42    M3 =    -2.30
```

TABLE CPUEB 45   (5 FOLDS)

```
L = 14.33    F0 =     0.27
B =   6.96   F1 =     0.82
t =   0.10   F2 =    -1.89    M2 =    -5.87
d =   0.96   F3 =     2.35    M3 =    -2.51
```

TABLE CPUEB 46   (5 FOLDS)

```
L = 14.33    F0 =     0.24
B =   7.17   F1 =     0.82
t =   0.10   F2 =    -1.84    M2 =    -6.15
d =   0.96   F3 =     2.29    M3 =    -2.73
```

TABLE CPUEB 47   (5 FOLDS)

```
L = 14.33    F0 =     0.21
B =   7.39   F1 =     0.82
t =   0.10   F2 =    -1.80    M2 =    -6.46
d =   0.96   F3 =     2.22    M3 =    -2.97
```

TABLE CPUEB 48   (5 FOLDS)

```
L = 14.33    F0 =     0.18
B =   7.61   F1 =     0.82
t =   0.10   F2 =    -1.75    M2 =    -6.77
d =   0.96   F3 =     2.16    M3 =    -3.21
```

TABLE CPUEB 49   (5 FOLDS)
```
L = 14.33    F0 =    0.15
B =  7.84    F1 =    0.82
t =  0.10    F2 =   -1.71    M2 =  -7.10
d =  0.96    F3 =    2.10    M3 =  -3.47
```

TABLE CPUEB 50   (5 FOLDS)
```
L = 14.76    F0 =    0.49
B =  6.75    F1 =    0.83
t =  0.12    F2 =   -1.97    M2 =  -6.51
d =  0.98    F3 =    2.45    M3 =  -2.44
```

TABLE CPUEB 51   (5 FOLDS)
```
L = 14.76    F0 =    0.47
B =  6.95    F1 =    0.83
t =  0.12    F2 =   -1.92    M2 =  -6.82
d =  0.98    F3 =    2.38    M3 =  -2.69
```

TABLE CPUEB 52   (5 FOLDS)
```
L = 14.76    F0 =    0.44
B =  7.16    F1 =    0.83
t =  0.12    F2 =   -1.87    M2 =  -7.16
d =  0.98    F3 =    2.32    M3 =  -2.96
```

TABLE CPUEB 53   (5 FOLDS)
```
L = 14.76    F0 =    0.41
B =  7.37    F1 =    0.83
t =  0.12    F2 =   -1.82    M2 =  -7.50
d =  0.98    F3 =    2.26    M3 =  -3.22
```

TABLE CPUEB 54   (5 FOLDS)
```
L = 14.76    F0 =    0.39
B =  7.59    F1 =    0.83
t =  0.12    F2 =   -1.78    M2 =  -7.87
d =  0.98    F3 =    2.20    M3 =  -3.50
```

TABLE CPUEB 55   (5 FOLDS)

```
L = 14.76   F0 =    0.36
B =  7.82   F1 =    0.82
t =  0.12   F2 =   -1.73   M2 =  -8.26
d =  0.98   F3 =    2.13   M3 =  -3.80
```

TABLE CPUEB 56   (5 FOLDS)

```
L = 14.76   F0 =    0.33
B =  8.05   F1 =    0.82
t =  0.12   F2 =   -1.69   M2 =  -8.67
d =  0.98   F3 =    2.08   M3 =  -4.11
```

TABLE CPUEB 57   (5 FOLDS)

```
L = 15.20   F0 =    0.51
B =  6.95   F1 =    0.86
t =  0.12   F2 =   -2.03   M2 =  -6.86
d =  1.01   F3 =    2.52   M3 =  -2.63
```

TABLE CPUEB 58   (5 FOLDS)

```
L = 15.20   F0 =    0.48
B =  7.16   F1 =    0.86
t =  0.12   F2 =   -1.97   M2 =  -7.20
d =  1.01   F3 =    2.46   M3 =  -2.90
```

TABLE CPUEB 59   (5 FOLDS)

```
L = 15.20   F0 =    0.45
B =  7.37   F1 =    0.85
t =  0.12   F2 =   -1.93   M2 =  -7.55
d =  1.01   F3 =    2.39   M3 =  -3.17
```

TABLE CPUEB 60   (5 FOLDS)

```
L = 15.20   F0 =    0.42
B =  7.59   F1 =    0.85
t =  0.12   F2 =   -1.88   M2 =  -7.92
d =  1.01   F3 =    2.32   M3 =  -3.45
```

TABLE CPUEB 61 (5 FOLDS)
L = 15.20   F0 =   0.40
B = 7.82   F1 =   0.85
t = 0.12   F2 = -1.83   M2 = -8.31
d = 1.01   F3 =   2.26   M3 = -3.75

TABLE CPUEB 62 (5 FOLDS)
L = 15.20   F0 =   0.37
B = 8.05   F1 =   0.85
t = 0.12   F2 = -1.78   M2 = -8.72
d = 1.01   F3 =   2.20   M3 = -4.06

TABLE CPUEB 63 (5 FOLDS)
L = 15.20   F0 =   0.34
B = 8.29   F1 =   0.85
t = 0.12   F2 = -1.74   M2 = -9.15
d = 1.01   F3 =   2.14   M3 = -4.38

TABLE CPUEB 64 (5 FOLDS)
L = 15.66   F0 =   0.52
B = 7.16   F1 =   0.88
t = 0.12   F2 = -2.09   M2 = -7.24
d = 1.04   F3 =   2.60   M3 = -2.84

TABLE CPUEB 65 (5 FOLDS)
L = 15.66   F0 =   0.49
B = 7.37   F1 =   0.88
t = 0.12   F2 = -2.04   M2 = -7.59
d = 1.04   F3 =   2.53   M3 = -3.11

TABLE CPUEB 66 (5 FOLDS)
L = 15.66   F0 =   0.47
B = 7.59   F1 =   0.88
t = 0.12   F2 = -1.98   M2 = -7.96
d = 1.04   F3 =   2.46   M3 = -3.40

TABLE CPUEB 67   (5 FOLDS)
L = 15.66   F0 =   0.44
B =   7.82   F1 =   0.88
t =   0.12   F2 =   -1.93   M2 =   -8.36
d =   1.04   F3 =   2.40   M3 =   -3.70

TABLE CPUEB 68   (5 FOLDS)
L = 15.66   F0 =   0.41
B =   8.05   F1 =   0.88
t =   0.12   F2 =   -1.88   M2 =   -8.77
d =   1.04   F3 =   2.33   M3 =   -4.01

TABLE CPUEB 69   (5 FOLDS)
L = 15.66   F0 =   0.38
B =   8.29   F1 =   0.87
t =   0.12   F2 =   -1.84   M2 =   -9.20
d =   1.04   F3 =   2.27   M3 =   -4.33

TABLE CPUEB 70   (5 FOLDS)
L = 15.66   F0 =   0.35
B =   8.54   F1 =   0.87
t =   0.12   F2 =   -1.79   M2 =   -9.67
d =   1.04   F3 =   2.20   M3 =   -4.68

TABLE CPUEB 71   (5 FOLDS)
L = 16.13   F0 =   0.53
B =   7.38   F1 =   0.91
t =   0.12   F2 =   -2.15   M2 =   -7.65
d =   1.08   F3 =   2.68   M3 =   -3.06

TABLE CPUEB 72   (5 FOLDS)
L = 16.13   F0 =   0.50
B =   7.60   F1 =   0.91
t =   0.12   F2 =   -2.09   M2 =   -8.02
d =   1.08   F3 =   2.61   M3 =   -3.35

TABLE CPUEB 73   (5 FOLDS)
```
L = 16.13    F0 =    0.47
B =  7.83    F1 =    0.90
t =  0.12    F2 =   -2.04    M2 =   -8.43
d =  1.08    F3 =    2.53    M3 =   -3.65
```

TABLE CPUEB 74   (5 FOLDS)
```
L = 16.13    F0 =    0.44
B =  8.06    F1 =    0.90
t =  0.12    F2 =   -1.99    M2 =   -8.84
d =  1.08    F3 =    2.47    M3 =   -3.97
```

TABLE CPUEB 75   (5 FOLDS)
```
L = 16.13    F0 =    0.42
B =  8.30    F1 =    0.90
t =  0.12    F2 =   -1.94    M2 =   -9.28
d =  1.08    F3 =    2.40    M3 =   -4.30
```

TABLE CPUEB 76   (5 FOLDS)
```
L = 16.13    F0 =    0.38
B =  8.55    F1 =    0.90
t =  0.12    F2 =   -1.89    M2 =   -9.75
d =  1.08    F3 =    2.33    M3 =   -4.64
```

TABLE CPUEB 77   (5 FOLDS)
```
L = 16.13    F0 =    0.35
B =  8.81    F1 =    0.90
t =  0.12    F2 =   -1.84    M2 = -10.25
d =  1.08    F3 =    2.27    M3 =   -5.02
```

TABLE CPUEB 78   (5 FOLDS)
```
L = 16.61    F0 =    0.55
B =  7.60    F1 =    0.94
t =  0.12    F2 =   -2.21    M2 =   -8.07
d =  1.11    F3 =    2.76    M3 =   -3.28
```

TABLE CPUEB 79   (5 FOLDS)
L = 16.61   F0 =   0.52
B =   7.83   F1 =   0.93
t =   0.12   F2 =   -2.16   M2 =   -8.47
d =   1.11   F3 =   2.68   M3 =   -3.60

TABLE CPUEB 80   (5 FOLDS)
L = 16.61   F0 =   0.49
B =   8.06   F1 =   0.93
t =   0.12   F2 =   -2.10   M2 =   -8.88
d =   1.11   F3 =   2.61   M3 =   -3.91

TABLE CPUEB 81   (5 FOLDS)
L = 16.61   F0 =   0.46
B =   8.30   F1 =   0.93
t =   0.12   F2 =   -2.05   M2 =   -9.33
d =   1.11   F3 =   2.54   M3 =   -4.24

TABLE CPUEB 82   (5 FOLDS)
L = 16.61   F0 =   0.43
B =   8.55   F1 =   0.93
t =   0.12   F2 =   -2.00   M2 =   -9.80
d =   1.11   F3 =   2.47   M3 =   -4.59

TABLE CPUEB 83   (5 FOLDS)
L = 16.61   F0 =   0.40
B =   8.81   F1 =   0.92
t =   0.12   F2 =   -1.94   M2 = -10.30
d =   1.11   F3 =   2.40   M3 =   -4.97

TABLE CPUEB 84   (5 FOLDS)
L = 16.61   F0 =   0.36
B =   9.07   F1 =   0.92
t =   0.12   F2 =   -1.90   M2 = -10.82
d =   1.11   F3 =   2.34   M3 =   -5.35

TABLE CPUEB 85   (5 FOLDS)
L = 17.11   F0 =   0.56
B =   7.83   F1 =   0.97
t =   0.12   F2 =   -2.28   M2 =   -8.51
d =   1.14   F3 =   2.84   M3 =   -3.53

TABLE CPUEB 86   (5 FOLDS)
L = 17.11   F0 =   0.54
B =   8.06   F1 =   0.96
t =   0.12   F2 =   -2.22   M2 =   -8.93
d =   1.14   F3 =   2.77   M3 =   -3.85

TABLE CPUEB 87   (5 FOLDS)
L = 17.11   F0 =   0.50
B =   8.30   F1 =   0.96
t =   0.12   F2 =   -2.17   M2 =   -9.37
d =   1.14   F3 =   2.69   M3 =   -4.19

TABLE CPUEB 88   (5 FOLDS)
L = 17.11   F0 =   0.47
B =   8.55   F1 =   0.96
t =   0.12   F2 =   -2.11   M2 =   -9.85
d =   1.14   F3 =   2.62   M3 =   -4.54

TABLE CPUEB 89   (5 FOLDS)
L = 17.11   F0 =   0.44
B =   8.81   F1 =   0.96
t =   0.12   F2 =   -2.06   M2 = -10.35
d =   1.14   F3 =   2.54   M3 = -4.91

TABLE CPUEB 90   (5 FOLDS)
L = 17.11   F0 =   0.41
B =   9.07   F1 =   0.95
t =   0.12   F2 =   -2.00   M2 = -10.87
d =   1.14   F3 =   2.47   M3 =   -5.30

TABLE CPUEB 91   (5 FOLDS)
```
L =  17.11    F0 =    0.38
B =   9.34    F1 =    0.95
t =   0.12    F2 =   -1.95    M2 = -11.43
d =   1.14    F3 =    2.41    M3 =  -5.70
```

TABLE CPUEB 92   (5 FOLDS)
```
L =  17.62    F0 =    0.72
B =   8.06    F1 =    1.01
t =   0.14    F2 =   -2.28    M2 = -10.12
d =   1.17    F3 =    2.84    M3 =  -4.14
```

TABLE CPUEB 93   (5 FOLDS)
```
L =  17.62    F0 =    0.69
B =   8.30    F1 =    1.00
t =   0.14    F2 =   -2.22    M2 = -10.63
d =   1.17    F3 =    2.76    M3 =  -4.53
```

TABLE CPUEB 94   (5 FOLDS)
```
L =  17.62    F0 =    0.66
B =   8.55    F1 =    0.99
t =   0.14    F2 =   -2.17    M2 = -11.18
d =   1.17    F3 =    2.68    M3 =  -4.94
```

TABLE CPUEB 95   (5 FOLDS)
```
L =  17.62    F0 =    0.63
B =   8.81    F1 =    0.99
t =   0.14    F2 =   -2.11    M2 = -11.76
d =   1.17    F3 =    2.61    M3 =  -5.38
```

TABLE CPUEB 96   (5 FOLDS)
```
L =  17.62    F0 =    0.60
B =   9.07    F1 =    0.98
t =   0.14    F2 =   -2.06    M2 = -12.36
d =   1.17    F3 =    2.54    M3 =  -5.82
```

TABLE CPUEB 97 (5 FOLDS)
```
L =  17.62     F0 =     0.57
B =   9.34     F1 =     0.98
t =   0.14     F2 =    -2.00     M2 = -12.99
d =   1.17     F3 =     2.47     M3 =  -6.28
```

TABLE CPUEB 98 (5 FOLDS)
```
L =  17.62     F0 =     0.54
B =   9.62     F1 =     0.97
t =   0.14     F2 =    -1.95     M2 = -13.67
d =   1.17     F3 =     2.40     M3 =  -6.77
```

TABLE CPUEB 99 (5 FOLDS)
```
L =  18.15     F0 =     0.74
B =   8.30     F1 =     1.03
t =   0.14     F2 =    -2.35     M2 = -10.68
d =   1.21     F3 =     2.92     M3 =  -4.44
```

TABLE CPUEB 100 (5 FOLDS)
```
L =  18.15     F0 =     0.71
B =   8.55     F1 =     1.03
t =   0.14     F2 =    -2.29     M2 = -11.23
d =   1.21     F3 =     2.84     M3 =  -4.86
```

TABLE CPUEB 101 (5 FOLDS)
```
L =  18.15     F0 =     0.68
B =   8.81     F1 =     1.02
t =   0.14     F2 =    -2.23     M2 = -11.81
d =   1.21     F3 =     2.76     M3 =  -5.30
```

TABLE CPUEB 102 (5 FOLDS)
```
L =  18.15     F0 =     0.65
B =   9.07     F1 =     1.02
t =   0.14     F2 =    -2.17     M2 = -12.41
d =   1.21     F3 =     2.69     M3 =  -5.75
```

TABLE CPUEB 103   (5 FOLDS)

L = 18.15	F0 = 0.62	
B = 9.34	F1 = 1.01	
t = 0.14	F2 = -2.12	M2 = -13.05
d = 1.21	F3 = 2.61	M3 = -6.21

TABLE CPUEB 104   (5 FOLDS)

L = 18.15	F0 = 0.59	
B = 9.62	F1 = 1.00	
t = 0.14	F2 = -2.06	M2 = -13.73
d = 1.21	F3 = 2.54	M3 = -6.70

TABLE CPUEB 105   (5 FOLDS)

L = 18.15	F0 = 0.55	
B = 9.91	F1 = 1.00	
t = 0.14	F2 = -2.01	M2 = -14.45
d = 1.21	F3 = 2.47	M3 = -7.22

TABLE CPUEB 106   (5 FOLDS)

L = 18.69	F0 = 0.76	
B = 8.55	F1 = 1.06	
t = 0.14	F2 = -2.42	M2 = -11.28
d = 1.25	F3 = 3.01	M3 = -4.78

TABLE CPUEB 107   (5 FOLDS)

L = 18.69	F0 = 0.73	
B = 8.81	F1 = 1.06	
t = 0.14	F2 = -2.36	M2 = -11.87
d = 1.25	F3 = 2.92	M3 = -5.22

TABLE CPUEB 108   (5 FOLDS)

L = 18.69	F0 = 0.70	
B = 9.07	F1 = 1.05	
t = 0.14	F2 = -2.30	M2 = -12.47
d = 1.25	F3 = 2.84	M3 = -5.67

TABLE CPUEB 109   (5 FOLDS)
```
L = 18.69    F0 =    0.67
B =  9.34    F1 =    1.04
t =  0.14    F2 =  -2.24    M2 = -13.11
d =  1.25    F3 =    2.77   M3 =  -6.14
```

TABLE CPUEB 110   (5 FOLDS)
```
L = 18.69    F0 =    0.63
B =  9.62    F1 =    1.04
t =  0.14    F2 =  -2.18    M2 = -13.79
d =  1.25    F3 =    2.69   M3 =  -6.64
```

TABLE CPUEB 111   (5 FOLDS)
```
L = 18.69    F0 =    0.60
B =  9.91    F1 =    1.03
t =  0.14    F2 =  -2.12    M2 = -14.52
d =  1.25    F3 =    2.62   M3 =  -7.16
```

TABLE CPUEB 112   (5 FOLDS)
```
L = 18.69    F0 =    0.57
B = 10.21    F1 =    1.03
t =  0.14    F2 =  -2.07    M2 = -15.29
d =  1.25    F3 =    2.54   M3 =  -7.71
```

TABLE CPUEB 113   (5 FOLDS)
```
L = 19.25    F0 =    0.78
B =  8.81    F1 =    1.10
t =  0.14    F2 =  -2.49    M2 = -11.91
d =  1.28    F3 =    3.10   M3 =  -5.14
```

TABLE CPUEB 114   (5 FOLDS)
```
L = 19.25    F0 =    0.75
B =  9.07    F1 =    1.09
t =  0.14    F2 =  -2.43    M2 = -12.52
d =  1.28    F3 =    3.01   M3 =  -5.59
```

TABLE CPUEB 115   (5 FOLDS)
L = 19.25   FO =   0.72
B = 9.34   F1 =   1.08
t = 0.14   F2 =   -2.37   M2 = -13.16
d = 1.28   F3 =   2.93   M3 = -6.07

TABLE CPUEB 116   (5 FOLDS)
L = 19.25   FO =   0.69
B = 9.62   F1 =   1.08
t = 0.14   F2 =   -2.31   M2 = -13.84
d = 1.28   F3 =   2.85   M3 = -6.57

TABLE CPUEB 117   (5 FOLDS)
L = 19.25   FO =   0.66
B = 9.91   F1 =   1.07
t = 0.14   F2 =   -2.25   M2 = -14.57
d = 1.28   F3 =   2.77   M3 = -7.09

TABLE CPUEB 118   (5 FOLDS)
L = 19.25   FO =   0.62
B = 10.21   F1 =   1.07
t = 0.14   F2 =   -2.19   M2 = -15.34
d = 1.28   F3 =   2.69   M3 = -7.65

TABLE CPUEB 119   (5 FOLDS)
L = 19.25   FO =   0.59
B = 10.52   F1 =   1.06
t = 0.14   F2 =   -2.13   M2 = -16.16
d = 1.28   F3 =   2.62   M3 = -8.23

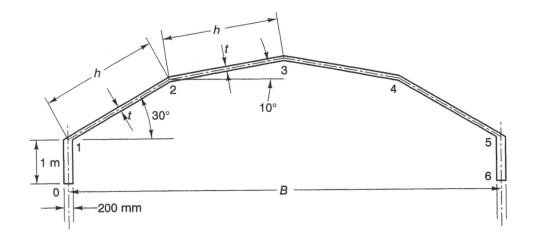

Fig. A5

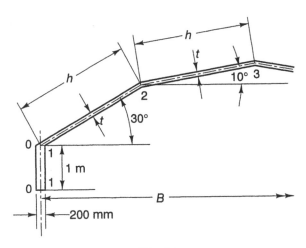

Fig. A6

Appendix 5. Design tables for folded plates types UDEB[1]

```
TABLE UDEB 1   (5 FOLDS)
L =  12.000    F0 =     1.90
B =   5.490    F1 =     0.10
t =   0.100    F2 =    -0.86    M2 =    -0.36
d =   1.000    F3 =    -1.09    M3 =    -1.63

TABLE UDEB 2   (5 FOLDS)
L =  12.000    F0 =     1.92
B =   5.650    F1 =     0.09
t =   0.100    F2 =    -0.84    M2 =    -0.40
d =   1.000    F3 =    -1.05    M3 =    -1.74

TABLE UDEB 3   (5 FOLDS)
L =  12.000    F0 =     1.95
B =   5.820    F1 =     0.08
t =   0.100    F2 =    -0.82    M2 =    -0.44
d =   1.000    F3 =    -1.01    M3 =    -1.86

TABLE UDEB 4   (5 FOLDS)
L =  12.000    F0 =     1.97
B =   5.990    F1 =     0.07
t =   0.100    F2 =    -0.81    M2 =    -0.47
d =   1.000    F3 =    -0.96    M3 =    -1.97

TABLE UDEB 5   (5 FOLDS)
L =  12.000    F0 =     2.00
B =   6.170    F1 =     0.05
t =   0.100    F2 =    -0.80    M2 =    -0.51
d =   1.000    F3 =    -0.91    M3 =    -2.08

TABLE UDEB 6   (5 FOLDS)
L =  12.000    F0 =     2.03
B =   6.360    F1 =     0.04
t =   0.100    F2 =    -0.79    M2 =    -0.54
d =   1.000    F3 =    -0.87    M3 =    -2.19
```

[1] See Figs A5 and A6

TABLE UDEB 7   (5 FOLDS)
L = 12.000   F0 =   2.06
B =   6.550   F1 =   0.02
t =   0.100   F2 =   −0.78   M2 =   −0.58
d =   1.000   F3 =   −0.82   M3 =   −2.30

TABLE UDEB 8   (5 FOLDS)
L = 12.360   F0 =   2.03
B =   5.650   F1 =   0.10
t =   0.100   F2 =   −0.89   M2 =   −0.42
d =   1.000   F3 =   −1.12   M3 =   −1.78

TABLE UDEB 9   (5 FOLDS)
L = 12.360   F0 =   2.06
B =   5.820   F1 =   0.09
t =   0.100   F2 =   −0.87   M2 =   −0.46
d =   1.000   F3 =   −1.07   M3 =   −1.90

TABLE UDEB 10   (5 FOLDS)
L = 12.360   F0 =   2.08
B =   5.990   F1 =   0.07
t =   0.100   F2 =   −0.86   M2 =   −0.50
d =   1.000   F3 =   −1.03   M3 =   −2.01

TABLE UDEB 11   (5 FOLDS)
L = 12.360   F0 =   2.11
B =   6.170   F1 =   0.06
t =   0.100   F2 =   −0.84   M2 =   −0.54
d =   1.000   F3 =   −0.98   M3 =   −2.13

TABLE UDEB 12   (5 FOLDS)
L = 12.360   F0 =   2.14
B =   6.360   F1 =   0.04
t =   0.100   F2 =   −0.83   M2 =   −0.57
d =   1.000   F3 =   −0.93   M3 =   −2.25

TABLE UDEB 13 (5 FOLDS)
L = 12.360   F0 =   2.18
B = 6.550   F1 =   0.03
t = 0.100   F2 = −0.82   M2 = −0.61
d = 1.000   F3 = −0.88   M3 = −2.36

TABLE UDEB 14 (5 FOLDS)
L = 12.360   F0 =   2.21
B = 6.750   F1 =   0.01
t = 0.100   F2 = −0.81   M2 = −0.64
d = 1.000   F3 = −0.83   M3 = −2.47

TABLE UDEB 15 (5 FOLDS)
L = 12.730   F0 =   2.18
B = 5.820   F1 =   0.10
t = 0.100   F2 = −0.92   M2 = −0.49
d = 1.000   F3 = −1.15   M3 = −1.94

TABLE UDEB 16 (5 FOLDS)
L = 12.730   F0 =   2.20
B = 5.990   F1 =   0.08
t = 0.100   F2 = −0.91   M2 = −0.53
d = 1.000   F3 = −1.10   M3 = −2.06

TABLE UDEB 17 (5 FOLDS)
L = 12.730   F0 =   2.23
B = 6.170   F1 =   0.07
t = 0.100   F2 = −0.89   M2 = −0.57
d = 1.000   F3 = −1.05   M3 = −2.18

TABLE UDEB 18 (5 FOLDS)
L = 12.730   F0 =   2.27
B = 6.360   F1 =   0.05
t = 0.100   F2 = −0.88   M2 = −0.61
d = 1.000   F3 = −1.00   M3 = −2.30

TABLE UDEB 19   (5 FOLDS)

L = 12.730	F0 =	2.30	
B = 6.550	F1 =	0.03	
t = 0.100	F2 =	-0.86	M2 = -0.64
d = 1.000	F3 =	-0.95	M3 = -2.42

TABLE UDEB 20   (5 FOLDS)

L = 12.730	F0 =	2.34	
B = 6.750	F1 =	0.02	
t = 0.100	F2 =	-0.85	M2 = -0.68
d = 1.000	F3 =	-0.89	M3 = -2.54

TABLE UDEB 21   (5 FOLDS)

L = 12.730	F0 =	2.37	
B = 6.950	F1 =	0.00	
t = 0.100	F2 =	-0.84	M2 = -0.72
d = 1.000	F3 =	-0.84	M3 = -2.65

TABLE UDEB 22   (5 FOLDS)

L = 13.110	F0 =	2.33	
B = 6.000	F1 =	0.10	
t = 0.100	F2 =	-0.96	M2 = -0.56
d = 1.000	F3 =	-1.17	M3 = -2.11

TABLE UDEB 23   (5 FOLDS)

L = 13.110	F0 =	2.36	
B = 6.180	F1 =	0.08	
t = 0.100	F2 =	-0.94	M2 = -0.60
d = 1.000	F3 =	-1.12	M3 = -2.24

TABLE UDEB 24   (5 FOLDS)

L = 13.110	F0 =	2.40	
B = 6.370	F1 =	0.06	
t = 0.100	F2 =	-0.92	M2 = -0.64
d = 1.000	F3 =	-1.07	M3 = -2.36

TABLE UDEB 25 (5 FOLDS)
L = 13.110   F0 =   2.43
B = 6.560   F1 =   0.04
t = 0.100   F2 = -0.91   M2 = -0.68
d = 1.000   F3 = -1.01   M3 = -2.49

TABLE UDEB 26 (5 FOLDS)
L = 13.110   F0 =   2.47
B = 6.760   F1 =   0.02
t = 0.100   F2 = -0.90   M2 = -0.72
d = 1.000   F3 = -0.96   M3 = -2.61

TABLE UDEB 27 (5 FOLDS)
L = 13.110   F0 =   2.51
B = 6.960   F1 =   0.00
t = 0.100   F2 = -0.89   M2 = -0.76
d = 1.000   F3 = -0.90   M3 = -2.72

TABLE UDEB 28 (5 FOLDS)
L = 13.110   F0 =   2.55
B = 7.170   F1 = -0.01
t = 0.100   F2 = -0.88   M2 = -0.79
d = 1.000   F3 = -0.85   M3 = -2.84

TABLE UDEB 29 (5 FOLDS)
L = 13.500   F0 =   2.49
B = 6.180   F1 =   0.09
t = 0.100   F2 = -0.99   M2 = -0.64
d = 1.000   F3 = -1.20   M3 = -2.29

TABLE UDEB 30 (5 FOLDS)
L = 13.500   F0 =   2.53
B = 6.370   F1 =   0.07
t = 0.100   F2 = -0.98   M2 = -0.68
d = 1.000   F3 = -1.14   M3 = -2.43

```
TABLE UDEB 31   (5 FOLDS)
L =  13.500    F0 =    2.57
B =   6.560    F1 =    0.05
t =   0.100    F2 =   -0.96    M2 =   -0.72
d =   1.000    F3 =   -1.09    M3 =   -2.55

TABLE UDEB 32   (5 FOLDS)
L =  13.500    F0 =    2.61
B =   6.760    F1 =    0.03
t =   0.100    F2 =   -0.95    M2 =   -0.76
d =   1.000    F3 =   -1.03    M3 =   -2.68

TABLE UDEB 33   (5 FOLDS)
L =  13.500    F0 =    2.65
B =   6.960    F1 =    0.01
t =   0.100    F2 =   -0.94    M2 =   -0.80
d =   1.000    F3 =   -0.97    M3 =   -2.80

TABLE UDEB 34   (5 FOLDS)
L =  13.500    F0 =    2.69
B =   7.170    F1 =   -0.01
t =   0.100    F2 =   -0.92    M2 =   -0.84
d =   1.000    F3 =   -0.91    M3 =   -2.92

TABLE UDEB 35   (5 FOLDS)
L =  13.500    F0 =    2.73
B =   7.390    F1 =   -0.03
t =   0.100    F2 =   -0.92    M2 =   -0.88
d =   1.000    F3 =   -0.85    M3 =   -3.04

TABLE UDEB 36   (5 FOLDS)
L =  13.910    F0 =    2.67
B =   6.360    F1 =    0.09
t =   0.100    F2 =   -1.04    M2 =   -0.72
d =   1.000    F3 =   -1.23    M3 =   -2.49
```

TABLE UDEB 37   (5 FOLDS)
L = 13.910   F0 =   2.71
B =   6.550   F1 =   0.07
t =   0.100   F2 =   −1.02   M2 =   −0.76
d =   1.000   F3 =   −1.17   M3 =   −2.62

TABLE UDEB 38   (5 FOLDS)
L = 13.910   F0 =   2.75
B =   6.750   F1 =   0.04
t =   0.100   F2 =   −1.00   M2 =   −0.81
d =   1.000   F3 =   −1.11   M3 =   −2.75

TABLE UDEB 39   (5 FOLDS)
L = 13.910   F0 =   2.80
B =   6.950   F1 =   0.02
t =   0.100   F2 =   −0.99   M2 =   −0.85
d =   1.000   F3 =   −1.05   M3 =   −2.88

TABLE UDEB 40   (5 FOLDS)
L = 13.910   F0 =   2.84
B =   7.160   F1 =   0.00
t =   0.100   F2 =   −0.98   M2 =   −0.89
d =   1.000   F3 =   −0.99   M3 =   −3.00

TABLE UDEB 41   (5 FOLDS)
L = 13.910   F0 =   2.89
B =   7.370   F1 =   −0.02
t =   0.100   F2 =   −0.97   M2 =   −0.92
d =   1.000   F3 =   −0.93   M3 =   −3.12

TABLE UDEB 42   (5 FOLDS)
L = 13.910   F0 =   2.93
B =   7.590   F1 =   −0.04
t =   0.100   F2 =   −0.96   M2 =   −0.97
d =   1.000   F3 =   −0.86   M3 =   −3.24

TABLE UDEB 43   (5 FOLDS)

L = 14.330	F0 =	2.86		
B = 6.560	F1 =	0.08		
t = 0.100	F2 =	-1.08	M2 =	-0.81
d = 1.000	F3 =	-1.25	M3 =	-2.70

TABLE UDEB 44   (5 FOLDS)

L = 14.330	F0 =	2.91		
B = 6.760	F1 =	0.06		
t = 0.100	F2 =	-1.06	M2 =	-0.86
d = 1.000	F3 =	-1.19	M3 =	-2.84

TABLE UDEB 45   (5 FOLDS)

L = 14.330	F0 =	2.96		
B = 6.960	F1 =	0.03		
t = 0.100	F2 =	-1.04	M2 =	-0.90
d = 1.000	F3 =	-1.12	M3 =	-2.97

TABLE UDEB 46   (5 FOLDS)

L = 14.330	F0 =	3.00		
B = 7.170	F1 =	0.01		
t = 0.100	F2 =	-1.03	M2 =	-0.94
d = 1.000	F3 =	-1.06	M3 =	-3.10

TABLE UDEB 47   (5 FOLDS)

L = 14.330	F0 =	3.05		
B = 7.390	F1 =	-0.02		
t = 0.100	F2 =	-1.02	M2 =	-0.98
d = 1.000	F3 =	-0.99	M3 =	-3.23

TABLE UDEB 48   (5 FOLDS)

L = 14.330	F0 =	3.10		
B = 7.610	F1 =	-0.04		
t = 0.100	F2 =	-1.01	M2 =	-1.02
d = 1.000	F3 =	-0.93	M3 =	-3.35

TABLE UDEB 49   (5 FOLDS)
```
L = 14.330    F0 =     3.16
B =  7.840    F1 =    -0.06
t =  0.100    F2 =    -1.00    M2 =    -1.07
d =  1.000    F3 =    -0.86    M3 =    -3.48
```

TABLE UDEB 50   (5 FOLDS)
```
L = 14.760    F0 =     3.07
B =  6.750    F1 =     0.07
t =  0.100    F2 =    -1.12    M2 =    -0.91
d =  1.000    F3 =    -1.28    M3 =    -2.92
```

TABLE UDEB 51   (5 FOLDS)
```
L = 14.760    F0 =     3.12
B =  6.950    F1 =     0.05
t =  0.100    F2 =    -1.10    M2 =    -0.95
d =  1.000    F3 =    -1.21    M3 =    -3.05
```

TABLE UDEB 52   (5 FOLDS)
```
L = 14.760    F0 =     3.17
B =  7.160    F1 =     0.02
t =  0.100    F2 =    -1.09    M2 =    -1.00
d =  1.000    F3 =    -1.14    M3 =    -3.19
```

TABLE UDEB 53   (5 FOLDS)
```
L = 14.760    F0 =     3.22
B =  7.370    F1 =     0.00
t =  0.100    F2 =    -1.07    M2 =    -1.04
d =  1.000    F3 =    -1.08    M3 =    -3.32
```

TABLE UDEB 54   (5 FOLDS)
```
L = 14.760    F0 =     3.27
B =  7.590    F1 =    -0.03
t =  0.100    F2 =    -1.06    M2 =    -1.08
d =  1.000    F3 =    -1.01    M3 =    -3.45
```

TABLE UDEB 55   (5 FOLDS)
```
L = 14.760    F0 =    3.33
B =  7.820    F1 =   -0.05
t =  0.100    F2 =   -1.05    M2 =   -1.12
d =  1.000    F3 =   -0.94    M3 =   -3.58
```

TABLE UDEB 56   (5 FOLDS)
```
L = 14.760    F0 =    3.38
B =  8.050    F1 =   -0.08
t =  0.100    F2 =   -1.05    M2 =   -1.17
d =  1.000    F3 =   -0.87    M3 =   -3.70
```

TABLE UDEB 57   (5 FOLDS)
```
L = 15.200    F0 =    3.28
B =  6.950    F1 =    0.07
t =  0.100    F2 =   -1.17    M2 =   -1.01
d =  1.000    F3 =   -1.30    M3 =   -3.15
```

TABLE UDEB 58   (5 FOLDS)
```
L = 15.200    F0 =    3.34
B =  7.160    F1 =    0.04
t =  0.100    F2 =   -1.15    M2 =   -1.06
d =  1.000    F3 =   -1.23    M3 =   -3.29
```

TABLE UDEB 59   (5 FOLDS)
```
L = 15.200    F0 =    3.39
B =  7.370    F1 =    0.01
t =  0.100    F2 =   -1.13    M2 =   -1.10
d =  1.000    F3 =   -1.16    M3 =   -3.43
```

TABLE UDEB 60   (5 FOLDS)
```
L = 15.200    F0 =    3.45
B =  7.590    F1 =   -0.02
t =  0.100    F2 =   -1.12    M2 =   -1.15
d =  1.000    F3 =   -1.09    M3 =   -3.56
```

TABLE UDEB 61   (5 FOLDS)

L = 15.200	F0 =	3.51		
B = 7.820	F1 =	-0.05		
t = 0.100	F2 =	-1.11	M2 =	-1.19
d = 1.000	F3 =	-1.02	M3 =	-3.70

TABLE UDEB 62   (5 FOLDS)

L = 15.200	F0 =	3.57		
B = 8.050	F1 =	-0.07		
t = 0.100	F2 =	-1.10	M2 =	-1.24
d = 1.000	F3 =	-0.95	M3 =	-3.82

TABLE UDEB 63   (5 FOLDS)

L = 15.200	F0 =	3.63		
B = 8.290	F1 =	-0.10		
t = 0.100	F2 =	-1.09	M2 =	-1.28
d = 1.000	F3 =	-0.87	M3 =	-3.95

TABLE UDEB 64   (5 FOLDS)

L = 15.660	F0 =	3.52		
B = 7.160	F1 =	0.06		
t = 0.100	F2 =	-1.22	M2 =	-1.13
d = 1.000	F3 =	-1.33	M3 =	-3.40

TABLE UDEB 65   (5 FOLDS)

L = 15.660	F0 =	3.58		
B = 7.370	F1 =	0.03		
t = 0.100	F2 =	-1.20	M2 =	-1.17
d = 1.000	F3 =	-1.25	M3 =	-3.55

TABLE UDEB 66   (5 FOLDS)

L = 15.660	F0 =	3.64		
B = 7.590	F1 =	0.00		
t = 0.100	F2 =	-1.18	M2 =	-1.22
d = 1.000	F3 =	-1.18	M3 =	-3.69

TABLE UDEB 67   (5 FOLDS)

L =	15.660	F0 =	3.71			
B =	7.820	F1 =	-0.03			
t =	0.100	F2 =	-1.17	M2 =	-1.27	
d =	1.000	F3 =	-1.10	M3 =	-3.82	

TABLE UDEB 68   (5 FOLDS)

L =	15.660	F0 =	3.77			
B =	8.050	F1 =	-0.06			
t =	0.100	F2 =	-1.16	M2 =	-1.31	
d =	1.000	F3 =	-1.03	M3 =	-3.96	

TABLE UDEB 69   (5 FOLDS)

L =	15.660	F0 =	3.84			
B =	8.290	F1 =	-0.09			
t =	0.100	F2 =	-1.15	M2 =	-1.36	
d =	1.000	F3 =	-0.95	M3 =	-4.09	

TABLE UDEB 70   (5 FOLDS)

L =	15.660	F0 =	3.90			
B =	8.540	F1 =	-0.12			
t =	0.100	F2 =	-1.15	M2 =	-1.41	
d =	1.000	F3 =	-0.87	M3 =	-4.22	

TABLE UDEB 71   (5 FOLDS)

L =	16.130	F0 =	3.78			
B =	7.380	F1 =	0.05			
t =	0.100	F2 =	-1.27	M2 =	-1.25	
d =	1.000	F3 =	-1.35	M3 =	-3.68	

TABLE UDEB 72   (5 FOLDS)

L =	16.130	F0 =	3.84			
B =	7.600	F1 =	0.01			
t =	0.100	F2 =	-1.25	M2 =	-1.30	
d =	1.000	F3 =	-1.27	M3 =	-3.82	

TABLE UDEB 73   (5 FOLDS)
```
L =  16.130    F0 =    3.91
B =   7.830    F1 =   -0.02
t =   0.100    F2 =   -1.23    M2 =   -1.35
d =   1.000    F3 =   -1.19    M3 =   -3.97
```

TABLE UDEB 74   (5 FOLDS)
```
L =  16.130    F0 =    3.98
B =   8.060    F1 =   -0.05
t =   0.100    F2 =   -1.22    M2 =   -1.39
d =   1.000    F3 =   -1.11    M3 =   -4.10
```

TABLE UDEB 75   (5 FOLDS)
```
L =  16.130    F0 =    4.05
B =   8.300    F1 =   -0.09
t =   0.100    F2 =   -1.21    M2 =   -1.44
d =   1.000    F3 =   -1.03    M3 =   -4.24
```

TABLE UDEB 76   (5 FOLDS)
```
L =  16.130    F0 =    4.12
B =   8.550    F1 =   -0.12
t =   0.100    F2 =   -1.20    M2 =   -1.49
d =   1.000    F3 =   -0.94    M3 =   -4.37
```

TABLE UDEB 77   (5 FOLDS)
```
L =  16.130    F0 =    4.20
B =   8.810    F1 =   -0.15
t =   0.100    F2 =   -1.20    M2 =   -1.54
d =   1.000    F3 =   -0.86    M3 =   -4.50
```

TABLE UDEB 78   (5 FOLDS)
```
L =  16.610    F0 =    4.04
B =   7.600    F1 =    0.03
t =   0.100    F2 =   -1.32    M2 =   -1.38
d =   1.000    F3 =   -1.37    M3 =   -3.96
```

TABLE UDEB 79   (5 FOLDS)
L = 16.610   F0 =   4.12
B =   7.830   F1 =   0.00
t =   0.100   F2 = -1.30   M2 = -1.43
d =   1.000   F3 = -1.28   M3 = -4.11

TABLE UDEB 80   (5 FOLDS)
L = 16.610   F0 =   4.19
B =   8.060   F1 = -0.04
t =   0.100   F2 = -1.29   M2 = -1.48
d =   1.000   F3 = -1.20   M3 = -4.25

TABLE UDEB 81   (5 FOLDS)
L = 16.610   F0 =   4.27
B =   8.300   F1 = -0.07
t =   0.100   F2 = -1.28   M2 = -1.53
d =   1.000   F3 = -1.11   M3 = -4.39

TABLE UDEB 82   (5 FOLDS)
L = 16.610   F0 =   4.35
B =   8.550   F1 = -0.11
t =   0.100   F2 = -1.27   M2 = -1.58
d =   1.000   F3 = -1.03   M3 = -4.53

TABLE UDEB 83   (5 FOLDS)
L = 16.610   F0 =   4.43
B =   8.810   F1 = -0.14
t =   0.100   F2 = -1.26   M2 = -1.63
d =   1.000   F3 = -0.94   M3 = -4.66

TABLE UDEB 84   (5 FOLDS)
L = 16.610   F0 =   4.51
B =   9.070   F1 = -0.17
t =   0.100   F2 = -1.26   M2 = -1.69
d =   1.000   F3 = -0.85   M3 = -4.79

TABLE UDEB 85   (5 FOLDS)
L = 17.110   F0 =   4.34
B =   7.830   F1 =   0.02
t =   0.100   F2 =   -1.38   M2 =   -1.53
d =   1.000   F3 =   -1.39   M3 =   -4.26

TABLE UDEB 86   (5 FOLDS)
L = 17.110   F0 =   4.41
B =   8.060   F1 =   -0.02
t =   0.100   F2 =   -1.36   M2 =   -1.58
d =   1.000   F3 =   -1.30   M3 =   -4.41

TABLE UDEB 87   (5 FOLDS)
L = 17.110   F0 =   4.50
B =   8.300   F1 =   -0.06
t =   0.100   F2 =   -1.35   M2 =   -1.63
d =   1.000   F3 =   -1.21   M3 =   -4.56

TABLE UDEB 88   (5 FOLDS)
L = 17.110   F0 =   4.58
B =   8.550   F1 =   -0.10
t =   0.100   F2 =   -1.34   M2 =   -1.68
d =   1.000   F3 =   -1.12   M3 =   -4.70

TABLE UDEB 89   (5 FOLDS)
L = 17.110   F0 =   4.67
B =   8.810   F1 =   -0.13
t =   0.100   F2 =   -1.33   M2 =   -1.73
d =   1.000   F3 =   -1.03   M3 =   -4.84

TABLE UDEB 90   (5 FOLDS)
L = 17.110   F0 =   4.75
B =   9.070   F1 =   -0.17
t =   0.100   F2 =   -1.32   M2 =   -1.79
d =   1.000   F3 =   -0.94   M3 =   -4.97

TABLE UDEB 91   (5 FOLDS)
```
L =  17.110    F0 =    4.84
B =   9.340    F1 =   -0.20
t =   0.100    F2 =   -1.32    M2 =   -1.85
d =   1.000    F3 =   -0.84    M3 =   -5.10
```

TABLE UDEB 92   (5 FOLDS)
```
L =  17.620    F0 =    4.64
B =   8.060    F1 =    0.01
t =   0.100    F2 =   -1.44    M2 =   -1.68
d =   1.000    F3 =   -1.41    M3 =   -4.58
```

TABLE UDEB 93   (5 FOLDS)
```
L =  17.620    F0 =    4.73
B =   8.300    F1 =   -0.04
t =   0.100    F2 =   -1.42    M2 =   -1.73
d =   1.000    F3 =   -1.31    M3 =   -4.73
```

TABLE UDEB 94   (5 FOLDS)
```
L =  17.620    F0 =    4.82
B =   8.550    F1 =   -0.08
t =   0.100    F2 =   -1.41    M2 =   -1.79
d =   1.000    F3 =   -1.21    M3 =   -4.88
```

TABLE UDEB 95   (5 FOLDS)
```
L =  17.620    F0 =    4.92
B =   8.810    F1 =   -0.12
t =   0.100    F2 =   -1.40    M2 =   -1.84
d =   1.000    F3 =   -1.12    M3 =   -5.03
```

TABLE UDEB 96   (5 FOLDS)
```
L =  17.620    F0 =    5.01
B =   9.070    F1 =   -0.16
t =   0.100    F2 =   -1.39    M2 =   -1.89
d =   1.000    F3 =   -1.02    M3 =   -5.16
```

TABLE UDEB 97   (5 FOLDS)
```
L = 17.620    F0 =    5.10
B =  9.340    F1 =   -0.19
t =  0.100    F2 =   -1.38    M2 =   -1.95
d =  1.000    F3 =   -0.93    M3 =   -5.30
```

TABLE UDEB 98   (5 FOLDS)
```
L = 17.620    F0 =    5.20
B =  9.620    F1 =   -0.23
t =  0.100    F2 =   -1.38    M2 =   -2.01
d =  1.000    F3 =   -0.83    M3 =   -5.43
```

TABLE UDEB 99   (5 FOLDS)
```
L = 18.150    F0 =    4.98
B =  8.300    F1 =   -0.01
t =  0.100    F2 =   -1.50    M2 =   -1.85
d =  1.000    F3 =   -1.42    M3 =   -4.92
```

TABLE UDEB 100   (5 FOLDS)
```
L = 18.150    F0 =    5.08
B =  8.550    F1 =   -0.06
t =  0.100    F2 =   -1.48    M2 =   -1.90
d =  1.000    F3 =   -1.32    M3 =   -5.08
```

TABLE UDEB 101   (5 FOLDS)
```
L = 18.150    F0 =    5.18
B =  8.810    F1 =   -0.10
t =  0.100    F2 =   -1.47    M2 =   -1.96
d =  1.000    F3 =   -1.22    M3 =   -5.23
```

TABLE UDEB 102   (5 FOLDS)
```
L = 18.150    F0 =    5.28
B =  9.070    F1 =   -0.15
t =  0.100    F2 =   -1.46    M2 =   -2.01
d =  1.000    F3 =   -1.12    M3 =   -5.37
```

```
TABLE UDEB 103   (5 FOLDS)
L = 18.150    F0 =    5.38
B =  9.340    F1 =   -0.19
t =  0.100    F2 =   -1.46    M2 =   -2.07
d =  1.000    F3 =   -1.02    M3 =   -5.51

TABLE UDEB 104   (5 FOLDS)
L = 18.150    F0 =    5.48
B =  9.620    F1 =   -0.22
t =  0.100    F2 =   -1.45    M2 =   -2.13
d =  1.000    F3 =   -0.91    M3 =   -5.65

TABLE UDEB 105   (5 FOLDS)
L = 18.150    F0 =    5.58
B =  9.910    F1 =   -0.26
t =  0.100    F2 =   -1.45    M2 =   -2.20
d =  1.000    F3 =   -0.81    M3 =   -5.78

TABLE UDEB 106   (5 FOLDS)
L = 18.690    F0 =    5.34
B =  8.550    F1 =   -0.03
t =  0.100    F2 =   -1.57    M2 =   -2.03
d =  1.000    F3 =   -1.43    M3 =   -5.29

TABLE UDEB 107   (5 FOLDS)
L = 18.690    F0 =    5.44
B =  8.810    F1 =   -0.08
t =  0.100    F2 =   -1.55    M2 =   -2.08
d =  1.000    F3 =   -1.32    M3 =   -5.45

TABLE UDEB 108   (5 FOLDS)
L = 18.690    F0 =    5.55
B =  9.070    F1 =   -0.13
t =  0.100    F2 =   -1.54    M2 =   -2.14
d =  1.000    F3 =   -1.22    M3 =   -5.59
```

TABLE UDEB 109   (5 FOLDS)
L = 18.690   F0 =   5.66
B =   9.340   F1 =   -0.17
t =   0.100   F2 =   -1.53   M2 =   -2.20
d =   1.000   F3 =   -1.11   M3 =   -5.74

TABLE UDEB 110   (5 FOLDS)
L = 18.690   F0 =   5.77
B =   9.620   F1 =   -0.22
t =   0.100   F2 =   -1.53   M2 =   -2.26
d =   1.000   F3 =   -1.00   M3 =   -5.88

TABLE UDEB 111   (5 FOLDS)
L = 18.690   F0 =   5.88
B =   9.910   F1 =   -0.26
t =   0.100   F2 =   -1.52   M2 =   -2.33
d =   1.000   F3 =   -0.90   M3 =   -6.01

TABLE UDEB 112   (5 FOLDS)
L = 18.690   F0 =   6.00
B = 10.210   F1 =   -0.30
t =   0.100   F2 =   -1.53   M2 =   -2.40
d =   1.000   F3 =   -0.79   M3 =   -6.14

TABLE UDEB 113   (5 FOLDS)
L = 19.250   F0 =   5.72
B =   8.810   F1 =   -0.05
t =   0.100   F2 =   -1.64   M2 =   -2.22
d =   1.000   F3 =   -1.44   M3 =   -5.68

TABLE UDEB 114   (5 FOLDS)
L = 19.250   F0 =   5.84
B =   9.070   F1 =   -0.10
t =   0.100   F2 =   -1.62   M2 =   -2.28
d =   1.000   F3 =   -1.33   M3 =   -5.83

```
TABLE UDEB 115   (5 FOLDS)
L = 19.250    F0 =    5.95
B =  9.340    F1 =   -0.15
t =  0.100    F2 =   -1.61    M2 =   -2.34
d =  1.000    F3 =   -1.22    M3 =   -5.98

TABLE UDEB 116   (5 FOLDS)
L = 19.250    F0 =    6.07
B =  9.620    F1 =   -0.20
t =  0.100    F2 =   -1.61    M2 =   -2.40
d =  1.000    F3 =   -1.10    M3 =   -6.12

TABLE UDEB 117   (5 FOLDS)
L = 19.250    F0 =    6.20
B =  9.910    F1 =   -0.25
t =  0.100    F2 =   -1.60    M2 =   -2.47
d =  1.000    F3 =   -0.99    M3 =   -6.26

TABLE UDEB 118   (5 FOLDS)
L = 19.250    F0 =    6.32
B = 10.210    F1 =   -0.29
t =  0.100    F2 =   -1.60    M2 =   -2.54
d =  1.000    F3 =   -0.87    M3 =   -6.40

TABLE UDEB 119   (5 FOLDS)
L = 19.250    F0 =    6.44
B = 10.520    F1 =   -0.33
t =  0.100    F2 =   -1.60    M2 =   -2.62
d =  1.000    F3 =   -0.76    M3 =   -6.52

TABLE UDEB 120   (5 FOLDS)
L = 19.830    F0 =    6.13
B =  9.070    F1 =   -0.07
t =  0.100    F2 =   -1.71    M2 =   -2.43
d =  1.000    F3 =   -1.45    M3 =   -6.09
```

TABLE UDEB 121 (5 FOLDS)
L = 19.830   F0 =   6.26
B = 9.340   F1 = -0.13
t = 0.100   F2 = -1.70   M2 = -2.49
d = 1.000   F3 = -1.33   M3 = -6.24

TABLE UDEB 122 (5 FOLDS)
L = 19.830   F0 =   6.39
B = 9.620   F1 = -0.18
t = 0.100   F2 = -1.69   M2 = -2.55
d = 1.000   F3 = -1.21   M3 = -6.39

TABLE UDEB 123 (5 FOLDS)
L = 19.830   F0 =   6.52
B = 9.910   F1 = -0.24
t = 0.100   F2 = -1.69   M2 = -2.62
d = 1.000   F3 = -1.09   M3 = -6.54

TABLE UDEB 124 (5 FOLDS)
L = 19.830   F0 =   6.66
B = 10.210   F1 = -0.29
t = 0.100   F2 = -1.68   M2 = -2.69
d = 1.000   F3 = -0.97   M3 = -6.67

TABLE UDEB 125 (5 FOLDS)
L = 19.830   F0 =   6.79
B = 10.520   F1 = -0.33
t = 0.100   F2 = -1.68   M2 = -2.77
d = 1.000   F3 = -0.85   M3 = -6.80

TABLE UDEB 126 (5 FOLDS)
L = 19.830   F0 =   6.92
B = 10.840   F1 = -0.37
t = 0.100   F2 = -1.69   M2 = -2.86
d = 1.000   F3 = -0.72   M3 = -6.93

TABLE UDEB 127   (5 FOLDS)

L = 20.420	F0 =	6.57		
B = 9.340	F1 =	-0.10		
t = 0.100	F2 =	-1.79	M2 =	-2.65
d = 1.000	F3 =	-1.46	M3 =	-6.52

TABLE UDEB 128   (5 FOLDS)

L = 20.420	F0 =	6.71		
B = 9.620	F1 =	-0.16		
t = 0.100	F2 =	-1.78	M2 =	-2.72
d = 1.000	F3 =	-1.33	M3 =	-6.68

TABLE UDEB 129   (5 FOLDS)

L = 20.420	F0 =	6.86		
B = 9.910	F1 =	-0.22		
t = 0.100	F2 =	-1.77	M2 =	-2.79
d = 1.000	F3 =	-1.20	M3 =	-6.83

TABLE UDEB 130   (5 FOLDS)

L = 20.420	F0 =	7.00		
B = 10.210	F1 =	-0.27		
t = 0.100	F2 =	-1.77	M2 =	-2.86
d = 1.000	F3 =	-1.07	M3 =	-6.97

TABLE UDEB 131   (5 FOLDS)

L = 20.420	F0 =	7.14		
B = 10.520	F1 =	-0.32		
t = 0.100	F2 =	-1.77	M2 =	-2.93
d = 1.000	F3 =	-0.94	M3 =	-7.10

TABLE UDEB 132   (5 FOLDS)

L = 20.420	F0 =	7.29		
B = 10.840	F1 =	-0.37		
t = 0.100	F2 =	-1.77	M2 =	-3.02
d = 1.000	F3 =	-0.81	M3 =	-7.23

TABLE UDEB 133   (5 FOLDS)
```
L = 20.420    F0 =    7.44
B = 11.170    F1 =   -0.42
t =  0.100    F2 =   -1.78    M2 =   -3.11
d =  1.000    F3 =   -0.68    M3 =   -7.35
```

TABLE UDEB 134   (5 FOLDS)
```
L = 21.030    F0 =    7.05
B =  9.620    F1 =   -0.12
t =  0.100    F2 =   -1.88    M2 =   -2.90
d =  1.000    F3 =   -1.45    M3 =   -6.98
```

TABLE UDEB 135   (5 FOLDS)
```
L = 21.030    F0 =    7.20
B =  9.910    F1 =   -0.19
t =  0.100    F2 =   -1.87    M2 =   -2.96
d =  1.000    F3 =   -1.32    M3 =   -7.14
```

TABLE UDEB 136   (5 FOLDS)
```
L = 21.030    F0 =    7.36
B = 10.210    F1 =   -0.25
t =  0.100    F2 =   -1.86    M2 =   -3.04
d =  1.000    F3 =   -1.18    M3 =   -7.29
```

TABLE UDEB 137   (5 FOLDS)
```
L = 21.030    F0 =    7.51
B = 10.520    F1 =   -0.31
t =  0.100    F2 =   -1.86    M2 =   -3.11
d =  1.000    F3 =   -1.04    M3 =   -7.43
```

TABLE UDEB 138   (5 FOLDS)
```
L = 21.030    F0 =    7.67
B = 10.840    F1 =   -0.36
t =  0.100    F2 =   -1.86    M2 =   -3.19
d =  1.000    F3 =   -0.91    M3 =   -7.56
```

TABLE UDEB 139 (5 FOLDS)
L = 21.030   F0 =   7.83
B = 11.170   F1 =   -0.41
t =   0.100   F2 =   -1.86   M2 =   -3.29
d =   1.000   F3 =   -0.77   M3 =   -7.68

TABLE UDEB 140 (5 FOLDS)
L = 21.030   F0 =   7.99
B = 11.510   F1 =   -0.46
t =   0.100   F2 =   -1.87   M2 =   -3.39
d =   1.000   F3 =   -0.64   M3 =   -7.80

Appendix 6. Design tables for folded plates types PDEB[1]

TABLE PDEB 1   (5 FOLDS)
```
L =  12.000    F0 =     0.58
B =   5.490    F1 =     1.01
t =   0.100    F2 =    -0.94    M2 =   -1.52
d =   1.000    F3 =    -1.28    M3 =   -3.41
```

TABLE PDEB 2   (5 FOLDS)
```
L =  12.000    F0 =     0.55
B =   5.650    F1 =     1.03
t =   0.100    F2 =    -0.93    M2 =   -1.57
d =   1.000    F3 =    -1.23    M3 =   -3.55
```

TABLE PDEB 3   (5 FOLDS)
```
L =  12.000    F0 =     0.52
B =   5.820    F1 =     1.04
t =   0.100    F2 =    -0.93    M2 =   -1.63
d =   1.000    F3 =    -1.18    M3 =   -3.70
```

TABLE PDEB 4   (5 FOLDS)
```
L =  12.000    F0 =     0.49
B =   5.990    F1 =     1.05
t =   0.100    F2 =    -0.92    M2 =   -1.68
d =   1.000    F3 =    -1.12    M3 =   -3.86
```

TABLE PDEB 5   (5 FOLDS)
```
L =  12.000    F0 =     0.46
B =   6.170    F1 =     1.07
t =   0.100    F2 =    -0.92    M2 =   -1.74
d =   1.000    F3 =    -1.06    M3 =   -4.02
```

TABLE PDEB 6   (5 FOLDS)
```
L =  12.000    F0 =     0.42
B =   6.360    F1 =     1.09
t =   0.100    F2 =    -0.92    M2 =   -1.80
d =   1.000    F3 =    -1.00    M3 =   -4.18
```

[1] See also Figs A5 and A6, p. 142

TABLE PDEB 7   (5 FOLDS)

```
L =  12.000    F0 =    0.38
B =   6.550    F1 =    1.11
t =   0.100    F2 =   -0.93    M2 =   -1.87
d =   1.000    F3 =   -0.94    M3 =   -4.34
```

TABLE PDEB 8   (5 FOLDS)

```
L =  12.360    F0 =    0.62
B =   5.650    F1 =    1.06
t =   0.100    F2 =   -0.97    M2 =   -1.59
d =   1.000    F3 =   -1.32    M3 =   -3.57
```

TABLE PDEB 9   (5 FOLDS)

```
L =  12.360    F0 =    0.59
B =   5.820    F1 =    1.08
t =   0.100    F2 =   -0.97    M2 =   -1.64
d =   1.000    F3 =   -1.26    M3 =   -3.73
```

TABLE PDEB 10   (5 FOLDS)

```
L =  12.360    F0 =    0.56
B =   5.990    F1 =    1.09
t =   0.100    F2 =   -0.96    M2 =   -1.70
d =   1.000    F3 =   -1.21    M3 =   -3.89
```

TABLE PDEB 11   (5 FOLDS)

```
L =  12.360    F0 =    0.52
B =   6.170    F1 =    1.11
t =   0.100    F2 =   -0.96    M2 =   -1.76
d =   1.000    F3 =   -1.15    M3 =   -4.05
```

TABLE PDEB 12   (5 FOLDS)

```
L =  12.360    F0 =    0.48
B =   6.360    F1 =    1.13
t =   0.100    F2 =   -0.96    M2 =   -1.82
d =   1.000    F3 =   -1.09    M3 =   -4.22
```

TABLE PDEB 13   (5 FOLDS)
```
L = 12.360    F0 =    0.45
B =  6.550    F1 =    1.15
t =  0.100    F2 =   -0.96    M2 =   -1.89
d =  1.000    F3 =   -1.02    M3 =   -4.39
```

TABLE PDEB 14   (5 FOLDS)
```
L = 12.360    F0 =    0.41
B =  6.750    F1 =    1.17
t =  0.100    F2 =   -0.97    M2 =   -1.95
d =  1.000    F3 =   -0.96    M3 =   -4.57
```

TABLE PDEB 15   (5 FOLDS)
```
L = 12.730    F0 =    0.65
B =  5.820    F1 =    1.12
t =  0.100    F2 =   -1.01    M2 =   -1.65
d =  1.000    F3 =   -1.36    M3 =   -3.75
```

TABLE PDEB 16   (5 FOLDS)
```
L = 12.730    F0 =    0.62
B =  5.990    F1 =    1.13
t =  0.100    F2 =   -1.00    M2 =   -1.71
d =  1.000    F3 =   -1.30    M3 =   -3.92
```

TABLE PDEB 17   (5 FOLDS)
```
L = 12.730    F0 =    0.59
B =  6.170    F1 =    1.15
t =  0.100    F2 =   -1.00    M2 =   -1.77
d =  1.000    F3 =   -1.24    M3 =   -4.09
```

TABLE PDEB 18   (5 FOLDS)
```
L = 12.730    F0 =    0.55
B =  6.360    F1 =    1.17
t =  0.100    F2 =   -0.99    M2 =   -1.84
d =  1.000    F3 =   -1.18    M3 =   -4.26
```

TABLE PDEB 19   (5 FOLDS)

L =	12.730	F0 =	0.51			
B =	6.550	F1 =	1.18			
t =	0.100	F2 =	-1.00	M2 =	-1.91
d =	1.000	F3 =	-1.11	M3 =	-4.44

TABLE PDEB 20   (5 FOLDS)

L =	12.730	F0 =	0.47			
B =	6.750	F1 =	1.21			
t =	0.100	F2 =	-1.00	M2 =	-1.97
d =	1.000	F3 =	-1.04	M3 =	-4.62

TABLE PDEB 21   (5 FOLDS)

L =	12.730	F0 =	0.43			
B =	6.950	F1 =	1.23			
t =	0.100	F2 =	-1.01	M2 =	-2.04
d =	1.000	F3 =	-0.97	M3 =	-4.80

TABLE PDEB 22   (5 FOLDS)

L =	13.110	F0 =	0.69			
B =	6.000	F1 =	1.18			
t =	0.100	F2 =	-1.04	M2 =	-1.73
d =	1.000	F3 =	-1.39	M3 =	-3.95

TABLE PDEB 23   (5 FOLDS)

L =	13.110	F0 =	0.66			
B =	6.180	F1 =	1.19			
t =	0.100	F2 =	-1.04	M2 =	-1.79
d =	1.000	F3 =	-1.33	M3 =	-4.12

TABLE PDEB 24   (5 FOLDS)

L =	13.110	F0 =	0.62			
B =	6.370	F1 =	1.21			
t =	0.100	F2 =	-1.03	M2 =	-1.86
d =	1.000	F3 =	-1.27	M3 =	-4.31

TABLE PDEB 25 (5 FOLDS)
L = 13.110   F0 =   0.58
B = 6.560   F1 =   1.23
t = 0.100   F2 =   -1.03   M2 =   -1.93
d = 1.000   F3 =   -1.20   M3 =   -4.49

TABLE PDEB 26 (5 FOLDS)
L = 13.110   F0 =   0.54
B = 6.760   F1 =   1.25
t = 0.100   F2 =   -1.04   M2 =   -2.00
d = 1.000   F3 =   -1.13   M3 =   -4.68

TABLE PDEB 27 (5 FOLDS)
L = 13.110   F0 =   0.50
B = 6.960   F1 =   1.27
t = 0.100   F2 =   -1.04   M2 =   -2.07
d = 1.000   F3 =   -1.06   M3 =   -4.86

TABLE PDEB 28 (5 FOLDS)
L = 13.110   F0 =   0.46
B = 7.170   F1 =   1.29
t = 0.100   F2 =   -1.06   M2 =   -2.14
d = 1.000   F3 =   -0.99   M3 =   -5.05

TABLE PDEB 29 (5 FOLDS)
L = 13.500   F0 =   0.73
B = 6.180   F1 =   1.24
t = 0.100   F2 =   -1.08   M2 =   -1.80
d = 1.000   F3 =   -1.43   M3 =   -4.15

TABLE PDEB 30 (5 FOLDS)
L = 13.500   F0 =   0.69
B = 6.370   F1 =   1.25
t = 0.100   F2 =   -1.08   M2 =   -1.87
d = 1.000   F3 =   -1.37   M3 =   -4.34

TABLE PDEB 31   (5 FOLDS)
```
L =  13.500     F0 =      0.66
B =   6.560     F1 =      1.27
t =   0.100     F2 =     -1.07     M2 =   -1.94
d =   1.000     F3 =     -1.30     M3 =   -4.53
```

TABLE PDEB 32   (5 FOLDS)
```
L =  13.500     F0 =      0.62
B =   6.760     F1 =      1.29
t =   0.100     F2 =     -1.07     M2 =   -2.01
d =   1.000     F3 =     -1.23     M3 =   -4.72
```

TABLE PDEB 33   (5 FOLDS)
```
L =  13.500     F0 =      0.58
B =   6.960     F1 =      1.31
t =   0.100     F2 =     -1.08     M2 =   -2.09
d =   1.000     F3 =     -1.16     M3 =   -4.91
```

TABLE PDEB 34   (5 FOLDS)
```
L =  13.500     F0 =      0.53
B =   7.170     F1 =      1.33
t =   0.100     F2 =     -1.09     M2 =   -2.16
d =   1.000     F3 =     -1.08     M3 =   -5.11
```

TABLE PDEB 35   (5 FOLDS)
```
L =  13.500     F0 =      0.48
B =   7.390     F1 =      1.36
t =   0.100     F2 =     -1.10     M2 =   -2.25
d =   1.000     F3 =     -1.00     M3 =   -5.31
```

TABLE PDEB 36   (5 FOLDS)
```
L =  13.910     F0 =      0.77
B =   6.360     F1 =      1.30
t =   0.100     F2 =     -1.12     M2 =   -1.88
d =   1.000     F3 =     -1.48     M3 =   -4.36
```

TABLE PDEB 37   (5 FOLDS)
```
L =  13.910    F0 =     0.74
B =   6.550    F1 =     1.32
t =   0.100    F2 =    -1.12    M2 =   -1.95
d =   1.000    F3 =    -1.41    M3 =   -4.55
```

TABLE PDEB 38   (5 FOLDS)
```
L =  13.910    F0 =     0.70
B =   6.750    F1 =     1.34
t =   0.100    F2 =    -1.11    M2 =   -2.03
d =   1.000    F3 =    -1.34    M3 =   -4.75
```

TABLE PDEB 39   (5 FOLDS)
```
L =  13.910    F0 =     0.66
B =   6.950    F1 =     1.35
t =   0.100    F2 =    -1.12    M2 =   -2.10
d =   1.000    F3 =    -1.26    M3 =   -4.95
```

TABLE PDEB 40   (5 FOLDS)
```
L =  13.910    F0 =     0.61
B =   7.160    F1 =     1.38
t =   0.100    F2 =    -1.12    M2 =   -2.18
d =   1.000    F3 =    -1.19    M3 =   -5.16
```

TABLE PDEB 41   (5 FOLDS)
```
L =  13.910    F0 =     0.57
B =   7.370    F1 =     1.40
t =   0.100    F2 =    -1.13    M2 =   -2.26
d =   1.000    F3 =    -1.11    M3 =   -5.36
```

TABLE PDEB 42   (5 FOLDS)
```
L =  13.910    F0 =     0.52
B =   7.590    F1 =     1.42
t =   0.100    F2 =    -1.15    M2 =   -2.34
d =   1.000    F3 =    -1.02    M3 =   -5.56
```

TABLE PDEB 43   (5 FOLDS)

L = 14.330	F0 =	0.82		
B = 6.560	F1 =	1.37		
t = 0.100	F2 =	-1.16	M2 =	-1.97
d = 1.000	F3 =	-1.51	M3 =	-4.59

TABLE PDEB 44   (5 FOLDS)

L = 14.330	F0 =	0.78		
B = 6.760	F1 =	1.39		
t = 0.100	F2 =	-1.16	M2 =	-2.04
d = 1.000	F3 =	-1.44	M3 =	-4.80

TABLE PDEB 45   (5 FOLDS)

L = 14.330	F0 =	0.74		
B = 6.960	F1 =	1.41		
t = 0.100	F2 =	-1.16	M2 =	-2.12
d = 1.000	F3 =	-1.37	M3 =	-5.00

TABLE PDEB 46   (5 FOLDS)

L = 14.330	F0 =	0.69		
B = 7.170	F1 =	1.43		
t = 0.100	F2 =	-1.16	M2 =	-2.20
d = 1.000	F3 =	-1.29	M3 =	-5.22

TABLE PDEB 47   (5 FOLDS)

L = 14.330	F0 =	0.65		
B = 7.390	F1 =	1.45		
t = 0.100	F2 =	-1.17	M2 =	-2.29
d = 1.000	F3 =	-1.21	M3 =	-5.44

TABLE PDEB 48   (5 FOLDS)

L = 14.330	F0 =	0.60		
B = 7.610	F1 =	1.47		
t = 0.100	F2 =	-1.18	M2 =	-2.37
d = 1.000	F3 =	-1.12	M3 =	-5.65

TABLE PDEB 49   (5 FOLDS)
L = 14.330   F0 =   0.55
B = 7.840   F1 =   1.50
t = 0.100   F2 = -1.20   M2 =   -2.46
d = 1.000   F3 = -1.03   M3 =   -5.86

TABLE PDEB 50   (5 FOLDS)
L = 14.760   F0 =   0.86
B = 6.750   F1 =   1.44
t = 0.100   F2 = -1.21   M2 =   -2.05
d = 1.000   F3 = -1.56   M3 =   -4.82

TABLE PDEB 51   (5 FOLDS)
L = 14.760   F0 =   0.82
B = 6.950   F1 =   1.46
t = 0.100   F2 = -1.20   M2 =   -2.13
d = 1.000   F3 = -1.48   M3 =   -5.03

TABLE PDEB 52   (5 FOLDS)
L = 14.760   F0 =   0.78
B = 7.160   F1 =   1.48
t = 0.100   F2 = -1.20   M2 =   -2.21
d = 1.000   F3 = -1.40   M3 =   -5.25

TABLE PDEB 53   (5 FOLDS)
L = 14.760   F0 =   0.74
B = 7.370   F1 =   1.50
t = 0.100   F2 = -1.21   M2 =   -2.30
d = 1.000   F3 = -1.32   M3 =   -5.47

TABLE PDEB 54   (5 FOLDS)
L = 14.760   F0 =   0.69
B = 7.590   F1 =   1.52
t = 0.100   F2 = -1.22   M2 =   -2.38
d = 1.000   F3 = -1.24   M3 =   -5.69

```
TABLE PDEB 55   (5 FOLDS)
L = 14.760    FO =    0.64
B =  7.820    F1 =    1.55
t =  0.100    F2 =  -1.23    M2 =   -2.48
d =  1.000    F3 =  -1.15    M3 =   -5.92

TABLE PDEB 56   (5 FOLDS)
L = 14.760    FO =    0.59
B =  8.050    F1 =    1.57
t =  0.100    F2 =  -1.25    M2 =   -2.57
d =  1.000    F3 =  -1.05    M3 =   -6.14

TABLE PDEB 57   (5 FOLDS)
L = 15.200    FO =    0.91
B =  6.950    F1 =    1.52
t =  0.100    F2 =  -1.25    M2 =   -2.14
d =  1.000    F3 =  -1.60    M3 =   -5.07

TABLE PDEB 58   (5 FOLDS)
L = 15.200    FO =    0.87
B =  7.160    F1 =    1.53
t =  0.100    F2 =  -1.25    M2 =   -2.23
d =  1.000    F3 =  -1.52    M3 =   -5.29

TABLE PDEB 59   (5 FOLDS)
L = 15.200    FO =    0.83
B =  7.370    F1 =    1.55
t =  0.100    F2 =  -1.25    M2 =   -2.31
d =  1.000    F3 =  -1.44    M3 =   -5.52

TABLE PDEB 60   (5 FOLDS)
L = 15.200    FO =    0.78
B =  7.590    F1 =    1.57
t =  0.100    F2 =  -1.26    M2 =   -2.40
d =  1.000    F3 =  -1.35    M3 =   -5.75
```

TABLE PDEB 61 (5 FOLDS)
```
L =  15.200    F0 =    0.73
B =   7.820    F1 =    1.60
t =   0.100    F2 =   -1.27    M2 =   -2.50
d =   1.000    F3 =   -1.26    M3 =   -5.99
```

TABLE PDEB 62 (5 FOLDS)
```
L =  15.200    F0 =    0.68
B =   8.050    F1 =    1.62
t =   0.100    F2 =   -1.29    M2 =   -2.59
d =   1.000    F3 =   -1.17    M3 =   -6.22
```

TABLE PDEB 63 (5 FOLDS)
```
L =  15.200    F0 =    0.63
B =   8.290    F1 =    1.65
t =   0.100    F2 =   -1.31    M2 =   -2.69
d =   1.000    F3 =   -1.07    M3 =   -6.45
```

TABLE PDEB 64 (5 FOLDS)
```
L =  15.660    F0 =    0.97
B =   7.160    F1 =    1.59
t =   0.100    F2 =   -1.30    M2 =   -2.24
d =   1.000    F3 =   -1.64    M3 =   -5.33
```

TABLE PDEB 65 (5 FOLDS)
```
L =  15.660    F0 =    0.92
B =   7.370    F1 =    1.61
t =   0.100    F2 =   -1.30    M2 =   -2.33
d =   1.000    F3 =   -1.56    M3 =   -5.56
```

TABLE PDEB 66 (5 FOLDS)
```
L =  15.660    F0 =    0.88
B =   7.590    F1 =    1.63
t =   0.100    F2 =   -1.30    M2 =   -2.42
d =   1.000    F3 =   -1.47    M3 =   -5.80
```

TABLE PDEB 67   (5 FOLDS)

```
L = 15.660    F0 =    0.83
B =  7.820    F1 =    1.65
t =  0.100    F2 =   -1.31    M2 =   -2.51
d =  1.000    F3 =   -1.38    M3 =   -6.05
```

TABLE PDEB 68   (5 FOLDS)

```
L = 15.660    F0 =    0.78
B =  8.050    F1 =    1.68
t =  0.100    F2 =   -1.33    M2 =   -2.61
d =  1.000    F3 =   -1.29    M3 =   -6.29
```

TABLE PDEB 69   (5 FOLDS)

```
L = 15.660    F0 =    0.72
B =  8.290    F1 =    1.71
t =  0.100    F2 =   -1.35    M2 =   -2.72
d =  1.000    F3 =   -1.18    M3 =   -6.53
```

TABLE PDEB 70   (5 FOLDS)

```
L = 15.660    F0 =    0.67
B =  8.540    F1 =    1.74
t =  0.100    F2 =   -1.37    M2 =   -2.82
d =  1.000    F3 =   -1.08    M3 =   -6.77
```

TABLE PDEB 71   (5 FOLDS)

```
L = 16.130    F0 =    1.02
B =  7.380    F1 =    1.68
t =  0.100    F2 =   -1.35    M2 =   -2.34
d =  1.000    F3 =   -1.68    M3 =   -5.61
```

TABLE PDEB 72   (5 FOLDS)

```
L = 16.130    F0 =    0.97
B =  7.600    F1 =    1.70
t =  0.100    F2 =   -1.35    M2 =   -2.44
d =  1.000    F3 =   -1.60    M3 =   -5.86
```

TABLE PDEB 73 (5 FOLDS)
L = 16.130   F0 =   0.93
B = 7.830   F1 =   1.72
t = 0.100   F2 = -1.36   M2 = -2.53
d = 1.000   F3 = -1.50   M3 = -6.11

TABLE PDEB 74 (5 FOLDS)
L = 16.130   F0 =   0.88
B = 8.060   F1 =   1.74
t = 0.100   F2 = -1.37   M2 = -2.63
d = 1.000   F3 = -1.41   M3 = -6.36

TABLE PDEB 75 (5 FOLDS)
L = 16.130   F0 =   0.82
B = 8.300   F1 =   1.77
t = 0.100   F2 = -1.38   M2 = -2.74
d = 1.000   F3 = -1.30   M3 = -6.62

TABLE PDEB 76 (5 FOLDS)
L = 16.130   F0 =   0.77
B = 8.550   F1 =   1.79
t = 0.100   F2 = -1.41   M2 = -2.85
d = 1.000   F3 = -1.20   M3 = -6.87

TABLE PDEB 77 (5 FOLDS)
L = 16.130   F0 =   0.71
B = 8.810   F1 =   1.83
t = 0.100   F2 = -1.44   M2 = -2.97
d = 1.000   F3 = -1.08   M3 = -7.13

TABLE PDEB 78 (5 FOLDS)
L = 16.610   F0 =   1.08
B = 7.600   F1 =   1.76
t = 0.100   F2 = -1.40   M2 = -2.45
d = 1.000   F3 = -1.73   M3 = -5.90

TABLE PDEB 79   (5 FOLDS)
```
L =  16.610     F0 =    1.03
B =   7.830     F1 =    1.78
t =   0.100     F2 =   -1.40     M2 =   -2.55
d =   1.000     F3 =   -1.63     M3 =   -6.16
```

TABLE PDEB 80   (5 FOLDS)
```
L =  16.610     F0 =    0.98
B =   8.060     F1 =    1.80
t =   0.100     F2 =   -1.41     M2 =   -2.65
d =   1.000     F3 =   -1.54     M3 =   -6.42
```

TABLE PDEB 81   (5 FOLDS)
```
L =  16.610     F0 =    0.93
B =   8.300     F1 =    1.83
t =   0.100     F2 =   -1.43     M2 =   -2.76
d =   1.000     F3 =   -1.43     M3 =   -6.69
```

TABLE PDEB 82   (5 FOLDS)
```
L =  16.610     F0 =    0.87
B =   8.550     F1 =    1.85
t =   0.100     F2 =   -1.44     M2 =   -2.87
d =   1.000     F3 =   -1.32     M3 =   -6.96
```

TABLE PDEB 83   (5 FOLDS)
```
L =  16.610     F0 =    0.81
B =   8.810     F1 =    1.89
t =   0.100     F2 =   -1.47     M2 =   -2.99
d =   1.000     F3 =   -1.21     M3 =   -7.23
```

TABLE PDEB 84   (5 FOLDS)
```
L =  16.610     F0 =    0.75
B =   9.070     F1 =    1.92
t =   0.100     F2 =   -1.50     M2 =   -3.11
d =   1.000     F3 =   -1.09     M3 =   -7.48
```

TABLE PDEB 85   (5 FOLDS)
```
L = 17.110    F0 =    1.14
B =  7.830    F1 =    1.85
t =  0.100    F2 =   -1.46    M2 =   -2.56
d =  1.000    F3 =   -1.77    M3 =   -6.21
```

TABLE PDEB 86   (5 FOLDS)
```
L = 17.110    F0 =    1.09
B =  8.060    F1 =    1.87
t =  0.100    F2 =   -1.46    M2 =   -2.66
d =  1.000    F3 =   -1.67    M3 =   -6.48
```

TABLE PDEB 87   (5 FOLDS)
```
L = 17.110    F0 =    1.04
B =  8.300    F1 =    1.90
t =  0.100    F2 =   -1.47    M2 =   -2.77
d =  1.000    F3 =   -1.57    M3 =   -6.76
```

TABLE PDEB 88   (5 FOLDS)
```
L = 17.110    F0 =    0.98
B =  8.550    F1 =    1.92
t =  0.100    F2 =   -1.49    M2 =   -2.89
d =  1.000    F3 =   -1.46    M3 =   -7.04
```

TABLE PDEB 89   (5 FOLDS)
```
L = 17.110    F0 =    0.92
B =  8.810    F1 =    1.95
t =  0.100    F2 =   -1.51    M2 =   -3.01
d =  1.000    F3 =   -1.34    M3 =   -7.32
```

TABLE PDEB 90   (5 FOLDS)
```
L = 17.110    F0 =    0.86
B =  9.070    F1 =    1.98
t =  0.100    F2 =   -1.54    M2 =   -3.13
d =  1.000    F3 =   -1.22    M3 =   -7.59
```

TABLE PDEB 91   (5 FOLDS)
```
L = 17.110    F0 =    0.80
B =  9.340    F1 =    2.02
t =  0.100    F2 =   -1.57    M2 =   -3.26
d =  1.000    F3 =   -1.10    M3 =   -7.85
```

TABLE PDEB 92   (5 FOLDS)
```
L = 17.620    F0 =    1.20
B =  8.060    F1 =    1.95
t =  0.100    F2 =   -1.51    M2 =   -2.68
d =  1.000    F3 =   -1.82    M3 =   -6.53
```

TABLE PDEB 93   (5 FOLDS)
```
L = 17.620    F0 =    1.15
B =  8.300    F1 =    1.97
t =  0.100    F2 =   -1.52    M2 =   -2.79
d =  1.000    F3 =   -1.71    M3 =   -6.81
```

TABLE PDEB 94   (5 FOLDS)
```
L = 17.620    F0 =    1.09
B =  8.550    F1 =    1.99
t =  0.100    F2 =   -1.53    M2 =   -2.90
d =  1.000    F3 =   -1.60    M3 =   -7.11
```

TABLE PDEB 95   (5 FOLDS)
```
L = 17.620    F0 =    1.03
B =  8.810    F1 =    2.02
t =  0.100    F2 =   -1.55    M2 =   -3.03
d =  1.000    F3 =   -1.48    M3 =   -7.40
```

TABLE PDEB 96   (5 FOLDS)
```
L = 17.620    F0 =    0.97
B =  9.070    F1 =    2.05
t =  0.100    F2 =   -1.58    M2 =   -3.15
d =  1.000    F3 =   -1.36    M3 =   -7.69
```

TABLE PDEB 97   (5 FOLDS)
L = 17.620   F0 =   0.91
B =   9.340   F1 =   2.08
t =   0.100   F2 =   -1.61   M2 =   -3.28
d =   1.000   F3 =   -1.23   M3 =   -7.97

TABLE PDEB 98   (5 FOLDS)
L = 17.620   F0 =   0.85
B =   9.620   F1 =   2.12
t =   0.100   F2 =   -1.65   M2 =   -3.42
d =   1.000   F3 =   -1.10   M3 =   -8.25

TABLE PDEB 99   (5 FOLDS)
L = 18.150   F0 =   1.27
B =   8.300   F1 =   2.05
t =   0.100   F2 =   -1.57   M2 =   -2.80
d =   1.000   F3 =   -1.86   M3 =   -6.87

TABLE PDEB 100   (5 FOLDS)
L = 18.150   F0 =   1.21
B =   8.550   F1 =   2.07
t =   0.100   F2 =   -1.58   M2 =   -2.92
d =   1.000   F3 =   -1.75   M3 =   -7.17

TABLE PDEB 101   (5 FOLDS)
L = 18.150   F0 =   1.15
B =   8.810   F1 =   2.09
t =   0.100   F2 =   -1.60   M2 =   -3.04
d =   1.000   F3 =   -1.63   M3 =   -7.48

TABLE PDEB 102   (5 FOLDS)
L = 18.150   F0 =   1.10
B =   9.070   F1 =   2.12
t =   0.100   F2 =   -1.62   M2 =   -3.17
d =   1.000   F3 =   -1.51   M3 =   -7.78

TABLE PDEB 103   (5 FOLDS)
```
L = 18.150    F0 =    1.03
B =  9.340    F1 =    2.15
t =  0.100    F2 =   -1.65    M2 =   -3.30
d =  1.000    F3 =   -1.38    M3 =   -8.08
```

TABLE PDEB 104   (5 FOLDS)
```
L = 18.150    F0 =    0.97
B =  9.620    F1 =    2.19
t =  0.100    F2 =   -1.68    M2 =   -3.44
d =  1.000    F3 =   -1.24    M3 =   -8.37
```

TABLE PDEB 105   (5 FOLDS)
```
L = 18.150    F0 =    0.90
B =  9.910    F1 =    2.22
t =  0.100    F2 =   -1.73    M2 =   -3.59
d =  1.000    F3 =   -1.10    M3 =   -8.66
```

TABLE PDEB 106   (5 FOLDS)
```
L = 18.690    F0 =    1.34
B =  8.550    F1 =    2.15
t =  0.100    F2 =   -1.64    M2 =   -2.93
d =  1.000    F3 =   -1.90    M3 =   -7.23
```

TABLE PDEB 107   (5 FOLDS)
```
L = 18.690    F0 =    1.28
B =  8.810    F1 =    2.18
t =  0.100    F2 =   -1.65    M2 =   -3.06
d =  1.000    F3 =   -1.78    M3 =   -7.55
```

TABLE PDEB 108   (5 FOLDS)
```
L = 18.690    F0 =    1.22
B =  9.070    F1 =    2.20
t =  0.100    F2 =   -1.67    M2 =   -3.18
d =  1.000    F3 =   -1.66    M3 =   -7.86
```

TABLE PDEB 109   (5 FOLDS)
```
L = 18.690    F0 =    1.16
B =  9.340    F1 =    2.23
t =  0.100    F2 =  -1.69    M2 =  -3.32
d =  1.000    F3 =  -1.53    M3 =  -8.18
```

TABLE PDEB 110   (5 FOLDS)
```
L = 18.690    F0 =    1.09
B =  9.620    F1 =    2.26
t =  0.100    F2 =  -1.72    M2 =  -3.46
d =  1.000    F3 =  -1.40    M3 =  -8.49
```

TABLE PDEB 111   (5 FOLDS)
```
L = 18.690    F0 =    1.02
B =  9.910    F1 =    2.30
t =  0.100    F2 =  -1.76    M2 =  -3.61
d =  1.000    F3 =  -1.25    M3 =  -8.80
```

TABLE PDEB 112   (5 FOLDS)
```
L = 18.690    F0 =    0.95
B = 10.210    F1 =    2.34
t =  0.100    F2 =  -1.81    M2 =  -3.77
d =  1.000    F3 =  -1.10    M3 =  -9.09
```

TABLE PDEB 113   (5 FOLDS)
```
L = 19.250    F0 =    1.41
B =  8.810    F1 =    2.26
t =  0.100    F2 =  -1.70    M2 =  -3.07
d =  1.000    F3 =  -1.95    M3 =  -7.61
```

TABLE PDEB 114   (5 FOLDS)
```
L = 19.250    F0 =    1.35
B =  9.070    F1 =    2.29
t =  0.100    F2 =  -1.72    M2 =  -3.20
d =  1.000    F3 =  -1.82    M3 =  -7.94
```

TABLE PDEB 115 (5 FOLDS)
```
L = 19.250    F0 =    1.29
B =  9.340    F1 =    2.31
t =  0.100    F2 =   -1.74    M2 =   -3.34
d =  1.000    F3 =   -1.69    M3 =   -8.26
```

TABLE PDEB 116 (5 FOLDS)
```
L = 19.250    F0 =    1.22
B =  9.620    F1 =    2.34
t =  0.100    F2 =   -1.77    M2 =   -3.48
d =  1.000    F3 =   -1.56    M3 =   -8.60
```

TABLE PDEB 117 (5 FOLDS)
```
L = 19.250    F0 =    1.16
B =  9.910    F1 =    2.38
t =  0.100    F2 =   -1.80    M2 =   -3.63
d =  1.000    F3 =   -1.41    M3 =   -8.92
```

TABLE PDEB 118 (5 FOLDS)
```
L = 19.250    F0 =    1.08
B = 10.210    F1 =    2.41
t =  0.100    F2 =   -1.85    M2 =   -3.79
d =  1.000    F3 =   -1.25    M3 =   -9.24
```

TABLE PDEB 119 (5 FOLDS)
```
L = 19.250    F0 =    1.01
B = 10.520    F1 =    2.46
t =  0.100    F2 =   -1.90    M2 =   -3.96
d =  1.000    F3 =   -1.09    M3 =   -9.54
```

TABLE PDEB 120 (5 FOLDS)
```
L = 19.830    F0 =    1.49
B =  9.070    F1 =    2.38
t =  0.100    F2 =   -1.77    M2 =   -3.21
d =  1.000    F3 =   -1.99    M3 =   -8.00
```

TABLE PDEB 121   (5 FOLDS)
L = 19.830   F0 =   1.43
B = 9.340   F1 =   2.40
t = 0.100   F2 = -1.79   M2 = -3.35
d = 1.000   F3 = -1.86   M3 = -8.35

TABLE PDEB 122   (5 FOLDS)
L = 19.830   F0 =   1.36
B = 9.620   F1 =   2.43
t = 0.100   F2 = -1.81   M2 = -3.50
d = 1.000   F3 = -1.72   M3 = -8.69

TABLE PDEB 123   (5 FOLDS)
L = 19.830   F0 =   1.29
B = 9.910   F1 =   2.46
t = 0.100   F2 = -1.85   M2 = -3.65
d = 1.000   F3 = -1.58   M3 = -9.04

TABLE PDEB 124   (5 FOLDS)
L = 19.830   F0 =   1.22
B = 10.210   F1 =   2.50
t = 0.100   F2 = -1.89   M2 = -3.81
d = 1.000   F3 = -1.42   M3 = -9.38

TABLE PDEB 125   (5 FOLDS)
L = 19.830   F0 =   1.15
B = 10.520   F1 =   2.54
t = 0.100   F2 = -1.94   M2 = -3.98
d = 1.000   F3 = -1.25   M3 = -9.71

TABLE PDEB 126   (5 FOLDS)
L = 19.830   F0 =   1.07
B = 10.840   F1 =   2.58
t = 0.100   F2 = -2.00   M2 = -4.17
d = 1.000   F3 = -1.08   M3 = -10.02

TABLE PDEB 127   (5 FOLDS)
L = 20.420   F0 =   1.57
B = 9.340   F1 =   2.50
t = 0.100   F2 = -1.85   M2 = -3.36
d = 1.000   F3 = -2.04   M3 = -8.42

TABLE PDEB 128   (5 FOLDS)
L = 20.420   F0 =   1.50
B = 9.620   F1 =   2.53
t = 0.100   F2 = -1.87   M2 = -3.51
d = 1.000   F3 = -1.90   M3 = -8.78

TABLE PDEB 129   (5 FOLDS)
L = 20.420   F0 =   1.44
B = 9.910   F1 =   2.56
t = 0.100   F2 = -1.89   M2 = -3.67
d = 1.000   F3 = -1.75   M3 = -9.14

TABLE PDEB 130   (5 FOLDS)
L = 20.420   F0 =   1.37
B = 10.210   F1 =   2.59
t = 0.100   F2 = -1.93   M2 = -3.83
d = 1.000   F3 = -1.59   M3 = -9.50

TABLE PDEB 131   (5 FOLDS)
L = 20.420   F0 =   1.29
B = 10.520   F1 =   2.63
t = 0.100   F2 = -1.98   M2 = -4.00
d = 1.000   F3 = -1.43   M3 = -9.86

TABLE PDEB 132   (5 FOLDS)
L = 20.420   F0 =   1.22
B = 10.840   F1 =   2.67
t = 0.100   F2 = -2.03   M2 = -4.19
d = 1.000   F3 = -1.25   M3 = -10.19

TABLE PDEB 133   (5 FOLDS)
```
L = 20.420    F0 =    1.14
B = 11.170    F1 =    2.71
t =  0.100    F2 =  -2.10    M2 =   -4.38
d =  1.000    F3 =  -1.07    M3 =  -10.51
```

TABLE PDEB 134   (5 FOLDS)
```
L = 21.030    F0 =    1.65
B =  9.620    F1 =    2.63
t =  0.100    F2 =  -1.92    M2 =   -3.52
d =  1.000    F3 =  -2.08    M3 =   -8.86
```

TABLE PDEB 135   (5 FOLDS)
```
L = 21.030    F0 =    1.59
B =  9.910    F1 =    2.65
t =  0.100    F2 =  -1.95    M2 =   -3.68
d =  1.000    F3 =  -1.93    M3 =   -9.24
```

TABLE PDEB 136   (5 FOLDS)
```
L = 21.030    F0 =    1.52
B = 10.210    F1 =    2.69
t =  0.100    F2 =  -1.98    M2 =   -3.85
d =  1.000    F3 =  -1.77    M3 =   -9.62
```

TABLE PDEB 137   (5 FOLDS)
```
L = 21.030    F0 =    1.44
B = 10.520    F1 =    2.72
t =  0.100    F2 =  -2.02    M2 =   -4.02
d =  1.000    F3 =  -1.61    M3 =   -9.99
```

TABLE PDEB 138   (5 FOLDS)
```
L = 21.030    F0 =    1.37
B = 10.840    F1 =    2.76
t =  0.100    F2 =  -2.07    M2 =   -4.20
d =  1.000    F3 =  -1.43    M3 =  -10.36
```

TABLE PDEB 139   (5 FOLDS)

L = 21.030	F0 = 1.29	
B = 11.170	F1 = 2.80	
t = 0.100	F2 = −2.13	M2 = −4.40
d = 1.000	F3 = −1.24	M3 = −10.70

TABLE PDEB 140   (5 FOLDS)

L = 21.030	F0 = 1.21	
B = 11.510	F1 = 2.85	
t = 0.100	F2 = −2.21	M2 = −4.61
d = 1.000	F3 = −1.05	M3 = −11.03

192

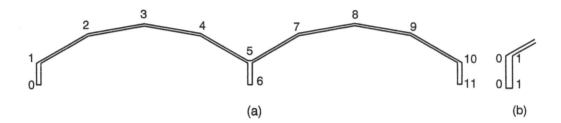

(a)

(b)

Fig. A7

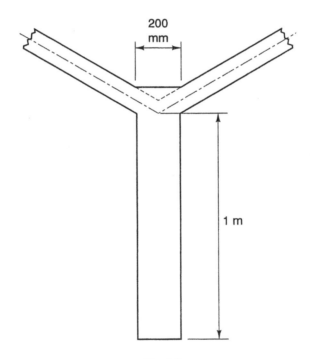

200
mm

1 m

Fig. A8

Appendix 7. Design tables for folded plates types CUDEB[1]

```
TABLE CUDEB 1   (9 FOLDS)
L =  12.000    F0 =     1.89
B =   5.490    F1 =     0.03
t =   0.100    F2 =    -0.69    M2 =    -0.62
d =   1.000    F3 =    -0.99    M3 =    -2.63
               F4 =    -1.19    M4 =    -1.84
               F5 =    -0.28    M5 =    -0.59
               F6 =     5.30


TABLE CUDEB 2   (9 FOLDS)
L =  12.000    F0 =     1.93
B =   5.650    F1 =     0.00
t =   0.100    F2 =    -0.67    M2 =    -0.61
d =   1.000    F3 =    -0.93    M3 =    -2.71
               F4 =    -1.18    M4 =    -1.93
               F5 =    -0.32    M5 =    -0.62
               F6 =     5.40


TABLE CUDEB 3   (9 FOLDS)
L =  12.000    F0 =     1.97
B =   5.820    F1 =    -0.03
t =   0.100    F2 =    -0.66    M2 =    -0.60
d =   1.000    F3 =    -0.87    M3 =    -2.79
               F4 =    -1.16    M4 =    -2.02
               F5 =    -0.36    M5 =    -0.66
               F6 =     5.51


TABLE CUDEB 4   (9 FOLDS)
L =  12.000    F0 =     2.01
B =   5.990    F1 =    -0.05
t =   0.100    F2 =    -0.65    M2 =    -0.58
d =   1.000    F3 =    -0.81    M3 =    -2.86
               F4 =    -1.14    M4 =    -2.10
               F5 =    -0.40    M5 =    -0.70
               F6 =     5.63
```

[1] See Figs A7 and A8

TABLE CUDEB 5   (9 FOLDS)

L = 12.000	F0 = 2.06		
B = 6.170	F1 = -0.08		
t = 0.100	F2 = -0.64	M2 = -0.56
d = 1.000	F3 = -0.75	M3 = -2.93
	F4 = -1.12	M4 = -2.19
	F5 = -0.44	M5 = -0.74
	F6 = 5.74		

TABLE CUDEB 6   (9 FOLDS)

L = 12.000	F0 = 2.10		
B = 6.360	F1 = -0.10		
t = 0.100	F2 = -0.64	M2 = -0.53
d = 1.000	F3 = -0.69	M3 = -3.00
	F4 = -1.10	M4 = -2.28
	F5 = -0.49	M5 = -0.79
	F6 = 5.87		

TABLE CUDEB 7   (9 FOLDS)

L = 12.000	F0 = 2.14		
B = 6.550	F1 = -0.13		
t = 0.100	F2 = -0.64	M2 = -0.51
d = 1.000	F3 = -0.63	M3 = -3.06
	F4 = -1.08	M4 = -2.37
	F5 = -0.54	M5 = -0.84
	F6 = 6.00		

TABLE CUDEB 8   (9 FOLDS)

L = 12.360	F0 = 2.03		
B = 5.650	F1 = 0.02		
t = 0.100	F2 = -0.71	M2 = -0.69
d = 1.000	F3 = -1.01	M3 = -2.84
	F4 = -1.25	M4 = -2.02
	F5 = -0.30	M5 = -0.62
	F6 = 5.69		

TABLE CUDEB 9   (9 FOLDS)

L = 12.360	F0 = 2.07	
B = 5.820	F1 = -0.01	
t = 0.100	F2 = -0.69	M2 = -0.68
d = 1.000	F3 = -0.94	M3 = -2.93
	F4 = -1.23	M4 = -2.11
	F5 = -0.35	M5 = -0.66
	F6 = 5.80	

TABLE CUDEB 10   (9 FOLDS)

L = 12.360	F0 = 2.12	
B = 5.990	F1 = -0.04	
t = 0.100	F2 = -0.68	M2 = -0.66
d = 1.000	F3 = -0.88	M3 = -3.01
	F4 = -1.22	M4 = -2.20
	F5 = -0.39	M5 = -0.70
	F6 = 5.92	

TABLE CUDEB 11   (9 FOLDS)

L = 12.360	F0 = 2.16	
B = 6.170	F1 = -0.07	
t = 0.100	F2 = -0.67	M2 = -0.64
d = 1.000	F3 = -0.81	M3 = -3.08
	F4 = -1.20	M4 = -2.29
	F5 = -0.43	M5 = -0.74
	F6 = 6.04	

TABLE CUDEB 12   (9 FOLDS)

L = 12.360	F0 = 2.21	
B = 6.360	F1 = -0.10	
t = 0.100	F2 = -0.67	M2 = -0.61
d = 1.000	F3 = -0.75	M3 = -3.15
	F4 = -1.18	M4 = -2.39
	F5 = -0.48	M5 = -0.79
	F6 = 6.18	

TABLE CUDEB 13 (9 FOLDS)

L = 12.360	F0 = 2.26		
B = 6.550	F1 = -0.13		
t = 0.100	F2 = -0.66	M2 = -0.59
d = 1.000	F3 = -0.68	M3 = -3.22
	F4 = -1.16	M4 = -2.48
	F5 = -0.53	M5 = -0.84
	F6 = 6.31		

TABLE CUDEB 14 (9 FOLDS)

L = 12.360	F0 = 2.30		
B = 6.750	F1 = -0.15		
t = 0.100	F2 = -0.67	M2 = -0.56
d = 1.000	F3 = -0.62	M3 = -3.28
	F4 = -1.14	M4 = -2.57
	F5 = -0.58	M5 = -0.89
	F6 = 6.45		

TABLE CUDEB 15 (9 FOLDS)

L = 12.730	F0 = 2.17		
B = 5.820	F1 = 0.01		
t = 0.100	F2 = -0.73	M2 = -0.77
d = 1.000	F3 = -1.02	M3 = -3.08
	F4 = -1.31	M4 = -2.21
	F5 = -0.33	M5 = -0.66
	F6 = 6.10		

TABLE CUDEB 16 (9 FOLDS)

L = 12.730	F0 = 2.22		
B = 5.990	F1 = -0.02		
t = 0.100	F2 = -0.71	M2 = -0.75
d = 1.000	F3 = -0.95	M3 = -3.16
	F4 = -1.30	M4 = -2.30
	F5 = -0.37	M5 = -0.70
	F6 = 6.22		

TABLE CUDEB 17   (9 FOLDS)

L = 12.730	F0 = 2.27	
B = 6.170	F1 = -0.06	
t = 0.100	F2 = -0.70	M2 = -0.73
d = 1.000	F3 = -0.89	M3 = -3.24
	F4 = -1.28	M4 = -2.40
	F5 = -0.42	M5 = -0.74
	F6 = 6.35	

TABLE CUDEB 18   (9 FOLDS)

L = 12.730	F0 = 2.32	
B = 6.360	F1 = -0.09	
t = 0.100	F2 = -0.69	M2 = -0.70
d = 1.000	F3 = -0.82	M3 = -3.32
	F4 = -1.26	M4 = -2.50
	F5 = -0.47	M5 = -0.79
	F6 = 6.50	

TABLE CUDEB 19   (9 FOLDS)

L = 12.730	F0 = 2.37	
B = 6.550	F1 = -0.12	
t = 0.100	F2 = -0.69	M2 = -0.67
d = 1.000	F3 = -0.75	M3 = -3.38
	F4 = -1.24	M4 = -2.60
	F5 = -0.52	M5 = -0.84
	F6 = 6.64	

TABLE CUDEB 20   (9 FOLDS)

L = 12.730	F0 = 2.42	
B = 6.750	F1 = -0.15	
t = 0.100	F2 = -0.69	M2 = -0.64
d = 1.000	F3 = -0.68	M3 = -3.45
	F4 = -1.22	M4 = -2.70
	F5 = -0.58	M5 = -0.89
	F6 = 6.79	

TABLE CUDEB 21   (9 FOLDS)

L =	12.730	F0 =	2.47		
B =	6.950	F1 =	-0.17		
t =	0.100	F2 =	-0.69	M2 =	-0.61
d =	1.000	F3 =	-0.61	M3 =	-3.50
		F4 =	-1.20	M4 =	-2.80
		F5 =	-0.63	M5 =	-0.94
		F6 =	6.94		

TABLE CUDEB 22   (9 FOLDS)

L =	13.110	F0 =	2.33		
B =	6.000	F1 =	-0.01		
t =	0.100	F2 =	-0.75	M2 =	-0.85
d =	1.000	F3 =	-1.03	M3 =	-3.33
		F4 =	-1.38	M4 =	-2.42
		F5 =	-0.36	M5 =	-0.70
		F6 =	6.55		

TABLE CUDEB 23   (9 FOLDS)

L =	13.110	F0 =	2.38		
B =	6.180	F1 =	-0.04		
t =	0.100	F2 =	-0.73	M2 =	-0.83
d =	1.000	F3 =	-0.96	M3 =	-3.42
		F4 =	-1.36	M4 =	-2.52
		F5 =	-0.41	M5 =	-0.74
		F6 =	6.69		

TABLE CUDEB 24   (9 FOLDS)

L =	13.110	F0 =	2.44		
B =	6.370	F1 =	-0.08		
t =	0.100	F2 =	-0.72	M2 =	-0.80
d =	1.000	F3 =	-0.89	M3 =	-3.50
		F4 =	-1.34	M4 =	-2.63
		F5 =	-0.46	M5 =	-0.79
		F6 =	6.83		

TABLE CUDEB 25   (9 FOLDS)
L = 13.110	F0 =	2.50			
B = 6.560	F1 =	-0.12			
t = 0.100	F2 =	-0.72	M2 =	-0.77	
d = 1.000	F3 =	-0.81	M3 =	-3.57	
	F4 =	-1.32	M4 =	-2.73	
	F5 =	-0.51	M5 =	-0.84	
	F6 =	6.98			

TABLE CUDEB 26   (9 FOLDS)
L = 13.110	F0 =	2.55			
B = 6.760	F1 =	-0.15			
t = 0.100	F2 =	-0.72	M2 =	-0.74	
d = 1.000	F3 =	-0.74	M3 =	-3.63	
	F4 =	-1.30	M4 =	-2.84	
	F5 =	-0.57	M5 =	-0.89	
	F6 =	7.14			

TABLE CUDEB 27   (9 FOLDS)
L = 13.110	F0 =	2.61			
B = 6.960	F1 =	-0.18			
t = 0.100	F2 =	-0.72	M2 =	-0.70	
d = 1.000	F3 =	-0.67	M3 =	-3.69	
	F4 =	-1.28	M4 =	-2.94	
	F5 =	-0.62	M5 =	-0.94	
	F6 =	7.30			

TABLE CUDEB 28   (9 FOLDS)
L = 13.110	F0 =	2.66			
B = 7.170	F1 =	-0.20			
t = 0.100	F2 =	-0.72	M2 =	-0.67	
d = 1.000	F3 =	-0.60	M3 =	-3.75	
	F4 =	-1.26	M4 =	-3.04	
	F5 =	-0.68	M5 =	-1.00	
	F6 =	7.47			

TABLE CUDEB 29   (9 FOLDS)

L = 13.500	F0 = 2.50		
B = 6.180	F1 = -0.02		
t = 0.100	F2 = -0.77	M2 = -0.93
d = 1.000	F3 = -1.05	M3 = -3.59
	F4 = -1.45	M4 = -2.64
	F5 = -0.38	M5 = -0.74
	F6 = 7.02		

TABLE CUDEB 30   (9 FOLDS)

L = 13.500	F0 = 2.56		
B = 6.370	F1 = -0.06		
t = 0.100	F2 = -0.76	M2 = -0.91
d = 1.000	F3 = -0.97	M3 = -3.68
	F4 = -1.43	M4 = -2.75
	F5 = -0.44	M5 = -0.79
	F6 = 7.18		

TABLE CUDEB 31   (9 FOLDS)

L = 13.500	F0 = 2.62		
B = 6.560	F1 = -0.10		
t = 0.100	F2 = -0.75	M2 = -0.88
d = 1.000	F3 = -0.89	M3 = -3.76
	F4 = -1.41	M4 = -2.86
	F5 = -0.50	M5 = -0.84
	F6 = 7.34		

TABLE CUDEB 32   (9 FOLDS)

L = 13.500	F0 = 2.68		
B = 6.760	F1 = -0.14		
t = 0.100	F2 = -0.74	M2 = -0.84
d = 1.000	F3 = -0.81	M3 = -3.83
	F4 = -1.39	M4 = -2.97
	F5 = -0.55	M5 = -0.89
	F6 = 7.50		

TABLE CUDEB 33   (9 FOLDS)

L = 13.500	F0 = 2.74		
B = 6.960	F1 = -0.17		
t = 0.100	F2 = -0.74	M2 = -0.80
d = 1.000	F3 = -0.74	M3 = -3.89
	F4 = -1.37	M4 = -3.08
	F5 = -0.61	M5 = -0.94
	F6 = 7.67		

TABLE CUDEB 34   (9 FOLDS)

L = 13.500	F0 = 2.80		
B = 7.170	F1 = -0.20		
t = 0.100	F2 = -0.75	M2 = -0.76
d = 1.000	F3 = -0.66	M3 = -3.95
	F4 = -1.35	M4 = -3.19
	F5 = -0.67	M5 = -1.00
	F6 = 7.85		

TABLE CUDEB 35   (9 FOLDS)

L = 13.500	F0 = 2.86		
B = 7.390	F1 = -0.23		
t = 0.100	F2 = -0.75	M2 = -0.73
d = 1.000	F3 = -0.58	M3 = -4.00
	F4 = -1.33	M4 = -3.31
	F5 = -0.73	M5 = -1.06
	F6 = 8.04		

TABLE CUDEB 36   (9 FOLDS)

L = 13.910	F0 = 2.68		
B = 6.360	F1 = -0.04		
t = 0.100	F2 = -0.80	M2 = -1.02
d = 1.000	F3 = -1.06	M3 = -3.88
	F4 = -1.53	M4 = -2.88
	F5 = -0.41	M5 = -0.79
	F6 = 7.53		

TABLE CUDEB 37   (9 FOLDS)

L = 13.910	F0 = 2.74	
B = 6.550	F1 = -0.08	
t = 0.100	F2 = -0.78	M2 = -1.00
d = 1.000	F3 = -0.98	M3 = -3.96
	F4 = -1.51	M4 = -2.99
	F5 = -0.47	M5 = -0.84
	F6 = 7.70	

TABLE CUDEB 38   (9 FOLDS)

L = 13.910	F0 = 2.81	
B = 6.750	F1 = -0.13	
t = 0.100	F2 = -0.78	M2 = -0.96
d = 1.000	F3 = -0.90	M3 = -4.04
	F4 = -1.49	M4 = -3.11
	F5 = -0.53	M5 = -0.89
	F6 = 7.88	

TABLE CUDEB 39   (9 FOLDS)

L = 13.910	F0 = 2.88	
B = 6.950	F1 = -0.17	
t = 0.100	F2 = -0.77	M2 = -0.92
d = 1.000	F3 = -0.82	M3 = -4.11
	F4 = -1.47	M4 = -3.23
	F5 = -0.59	M5 = -0.94
	F6 = 8.06	

TABLE CUDEB 40   (9 FOLDS)

L = 13.910	F0 = 2.94	
B = 7.160	F1 = -0.20	
t = 0.100	F2 = -0.77	M2 = -0.88
d = 1.000	F3 = -0.73	M3 = -4.17
	F4 = -1.44	M4 = -3.35
	F5 = -0.66	M5 = -1.00
	F6 = 8.24	

TABLE CUDEB 41   (9 FOLDS)

L = 13.910	F0 =	3.01			
B = 7.370	F1 =	-0.23			
t = 0.100	F2 =	-0.78	M2 =	-0.84
d = 1.000	F3 =	-0.66	M3 =	-4.22
	F4 =	-1.42	M4 =	-3.46
	F5 =	-0.72	M5 =	-1.06
	F6 =	8.43		

TABLE CUDEB 42   (9 FOLDS)

L = 13.910	F0 =	3.07		
B = 7.590	F1 =	-0.26		
t = 0.100	F2 =	-0.78	M2 =	-0.80
d = 1.000	F3 =	-0.58	M3 =	-4.27
	F4 =	-1.40	M4 =	-3.58
	F5 =	-0.78	M5 =	-1.12
	F6 =	8.63		

TABLE CUDEB 43   (9 FOLDS)

L = 14.330	F0 =	2.88		
B = 6.560	F1 =	-0.06		
t = 0.100	F2 =	-0.82	M2 =	-1.12
d = 1.000	F3 =	-1.07	M3 =	-4.18
	F4 =	-1.60	M4 =	-3.14
	F5 =	-0.45	M5 =	-0.84
	F6 =	8.10		

TABLE CUDEB 44   (9 FOLDS)

L = 14.330	F0 =	2.95		
B = 6.760	F1 =	-0.11		
t = 0.100	F2 =	-0.81	M2 =	-1.09
d = 1.000	F3 =	-0.98	M3 =	-4.27
	F4 =	-1.58	M4 =	-3.27
	F5 =	-0.51	M5 =	-0.89
	F6 =	8.28		

TABLE CUDEB 45   (9 FOLDS)

L = 14.330	F0 = 3.02
B = 6.960	F1 = -0.16
t = 0.100	F2 = -0.80 M2 = -1.05
d = 1.000	F3 = -0.89 M3 = -4.34
	F4 = -1.56 M4 = -3.39
	F5 = -0.58 M5 = -0.94
	F6 = 8.47

TABLE CUDEB 46   (9 FOLDS)

L = 14.330	F0 = 3.10
B = 7.170	F1 = -0.20
t = 0.100	F2 = -0.80 M2 = -1.00
d = 1.000	F3 = -0.81 M3 = -4.41
	F4 = -1.54 M4 = -3.52
	F5 = -0.64 M5 = -1.00
	F6 = 8.67

TABLE CUDEB 47   (9 FOLDS)

L = 14.330	F0 = 3.17
B = 7.390	F1 = -0.24
t = 0.100	F2 = -0.80 M2 = -0.95
d = 1.000	F3 = -0.72 M3 = -4.47
	F4 = -1.52 M4 = -3.65
	F5 = -0.71 M5 = -1.06
	F6 = 8.88

TABLE CUDEB 48   (9 FOLDS)

L = 14.330	F0 = 3.24
B = 7.610	F1 = -0.27
t = 0.100	F2 = -0.81 M2 = -0.91
d = 1.000	F3 = -0.63 M3 = -4.52
	F4 = -1.50 M4 = -3.77
	F5 = -0.78 M5 = -1.13
	F6 = 9.09

TABLE CUDEB 49   (9 FOLDS)

L = 14.330	F0 =	3.31		
B = 7.840	F1 =	-0.30		
t = 0.100	F2 =	-0.82	M2 =	-0.86
d = 1.000	F3 =	-0.55	M3 =	-4.57
	F4 =	-1.47	M4 =	-3.90
	F5 =	-0.85	M5 =	-1.20
	F6 =	9.31		

TABLE CUDEB 50   (9 FOLDS)

L = 14.760	F0 =	3.08		
B = 6.750	F1 =	-0.08		
t = 0.100	F2 =	-0.85	M2 =	-1.22
d = 1.000	F3 =	-1.08	M3 =	-4.50
	F4 =	-1.69	M4 =	-3.42
	F5 =	-0.48	M5 =	-0.89
	F6 =	8.68		

TABLE CUDEB 51   (9 FOLDS)

L = 14.760	F0 =	3.16		
B = 6.950	F1 =	-0.14		
t = 0.100	F2 =	-0.84	M2 =	-1.19
d = 1.000	F3 =	-0.99	M3 =	-4.59
	F4 =	-1.67	M4 =	-3.55
	F5 =	-0.54	M5 =	-0.94
	F6 =	8.88		

TABLE CUDEB 52   (9 FOLDS)

L = 14.760	F0 =	3.24		
B = 7.160	F1 =	-0.18		
t = 0.100	F2 =	-0.83	M2 =	-1.14
d = 1.000	F3 =	-0.89	M3 =	-4.66
	F4 =	-1.65	M4 =	-3.69
	F5 =	-0.62	M5 =	-1.00
	F6 =	9.09		

TABLE CUDEB 53   (9 FOLDS)

L = 14.760	F0 =	3.32		
B = 7.370	F1 =	-0.23		
t = 0.100	F2 =	-0.83	M2 =	-1.09
d = 1.000	F3 =	-0.80	M3 =	-4.72
	F4 =	-1.63	M4 =	-3.82
	F5 =	-0.69	M5 =	-1.06
	F6 =	9.30		

TABLE CUDEB 54   (9 FOLDS)

L = 14.760	F0 =	3.40		
B = 7.590	F1 =	-0.27		
t = 0.100	F2 =	-0.83	M2 =	-1.04
d = 1.000	F3 =	-0.71	M3 =	-4.78
	F4 =	-1.61	M4 =	-3.95
	F5 =	-0.76	M5 =	-1.12
	F6 =	9.52		

TABLE CUDEB 55   (9 FOLDS)

L = 14.760	F0 =	3.48		
B = 7.820	F1 =	-0.30		
t = 0.100	F2 =	-0.84	M2 =	-0.99
d = 1.000	F3 =	-0.62	M3 =	-4.83
	F4 =	-1.58	M4 =	-4.09
	F5 =	-0.83	M5 =	-1.19
	F6 =	9.76		

TABLE CUDEB 56   (9 FOLDS)

L = 14.760	F0 =	3.55		
B = 8.050	F1 =	-0.33		
t = 0.100	F2 =	-0.86	M2 =	-0.95
d = 1.000	F3 =	-0.54	M3 =	-4.87
	F4 =	-1.56	M4 =	-4.22
	F5 =	-0.91	M5 =	-1.26
	F6 =	9.99		

TABLE CUDEB 57   (9 FOLDS)

L = 15.200	F0 =	3.31		
B = 6.950	F1 =	-0.11		
t = 0.100	F2 =	-0.87	M2 =	-1.33
d = 1.000	F3 =	-1.08	M3 =	-4.84
	F4 =	-1.78	M4 =	-3.72
	F5 =	-0.51	M5 =	-0.94
	F6 =	9.31		

TABLE CUDEB 58   (9 FOLDS)

L = 15.200	F0 =	3.40		
B = 7.160	F1 =	-0.17		
t = 0.100	F2 =	-0.86	M2 =	-1.29
d = 1.000	F3 =	-0.98	M3 =	-4.93
	F4 =	-1.76	M4 =	-3.86
	F5 =	-0.59	M5 =	-1.00
	F6 =	9.53		

TABLE CUDEB 59   (9 FOLDS)

L = 15.200	F0 =	3.48		
B = 7.370	F1 =	-0.22		
t = 0.100	F2 =	-0.86	M2 =	-1.24
d = 1.000	F3 =	-0.89	M3 =	-5.00
	F4 =	-1.74	M4 =	-4.00
	F5 =	-0.66	M5 =	-1.06
	F6 =	9.76		

TABLE CUDEB 60   (9 FOLDS)

L = 15.200	F0 =	3.57		
B = 7.590	F1 =	-0.26		
t = 0.100	F2 =	-0.86	M2 =	-1.19
d = 1.000	F3 =	-0.79	M3 =	-5.06
	F4 =	-1.72	M4 =	-4.14
	F5 =	-0.74	M5 =	-1.12
	F6 =	9.99		

TABLE CUDEB 61   (9 FOLDS)

```
L = 15.200    F0 =     3.65
B =  7.820    F1 =    -0.31
t =  0.100    F2 =    -0.87    M2 =    -1.13
d =  1.000    F3 =    -0.69    M3 =    -5.11
              F4 =    -1.69    M4 =    -4.29
              F5 =    -0.81    M5 =    -1.19
              F6 =    10.24
```

TABLE CUDEB 62   (9 FOLDS)

```
L = 15.200    F0 =     3.74
B =  8.050    F1 =    -0.34
t =  0.100    F2 =    -0.88    M2 =    -1.08
d =  1.000    F3 =    -0.60    M3 =    -5.15
              F4 =    -1.67    M4 =    -4.43
              F5 =    -0.89    M5 =    -1.26
              F6 =    10.48
```

TABLE CUDEB 63   (9 FOLDS)

```
L = 15.200    F0 =     3.82
B =  8.290    F1 =    -0.37
t =  0.100    F2 =    -0.90    M2 =    -1.03
d =  1.000    F3 =    -0.51    M3 =    -5.19
              F4 =    -1.64    M4 =    -4.58
              F5 =    -0.97    M5 =    -1.34
              F6 =    10.74
```

TABLE CUDEB 64   (9 FOLDS)

```
L = 15.660    F0 =     3.55
B =  7.160    F1 =    -0.14
t =  0.100    F2 =    -0.90    M2 =    -1.45
d =  1.000    F3 =    -1.09    M3 =    -5.21
              F4 =    -1.88    M4 =    -4.05
              F5 =    -0.55    M5 =    -1.00
              F6 =    10.00
```

TABLE CUDEB 65 (9 FOLDS)

L = 15.660	F0 =	3.64		
B = 7.370	F1 =	-0.20		
t = 0.100	F2 =	-0.89	M2 =	-1.40
d = 1.000	F3 =	-0.98	M3 =	-5.29
	F4 =	-1.85	M4 =	-4.20
	F5 =	-0.63	M5 =	-1.06
	F6 =	10.24		

TABLE CUDEB 66 (9 FOLDS)

L = 15.660	F0 =	3.74		
B = 7.590	F1 =	-0.25		
t = 0.100	F2 =	-0.89	M2 =	-1.35
d = 1.000	F3 =	-0.88	M3 =	-5.36
	F4 =	-1.83	M4 =	-4.35
	F5 =	-0.71	M5 =	-1.12
	F6 =	10.48		

TABLE CUDEB 67 (9 FOLDS)

L = 15.660	F0 =	3.84		
B = 7.820	F1 =	-0.30		
t = 0.100	F2 =	-0.90	M2 =	-1.29
d = 1.000	F3 =	-0.78	M3 =	-5.42
	F4 =	-1.81	M4 =	-4.50
	F5 =	-0.79	M5 =	-1.19
	F6 =	10.74		

TABLE CUDEB 68 (9 FOLDS)

L = 15.660	F0 =	3.93		
B = 8.050	F1 =	-0.35		
t = 0.100	F2 =	-0.91	M2 =	-1.23
d = 1.000	F3 =	-0.68	M3 =	-5.46
	F4 =	-1.79	M4 =	-4.65
	F5 =	-0.87	M5 =	-1.26
	F6 =	11.00		

TABLE CUDEB 69   (9 FOLDS)

L = 15.660	F0 = 4.02		
B = 8.290	F1 = -0.38		
t = 0.100	F2 = -0.92	M2 =	-1.17
d = 1.000	F3 = -0.58	M3 =	-5.50
	F4 = -1.76	M4 =	-4.81
	F5 = -0.96	M5 =	-1.34
	F6 = 11.28		

TABLE CUDEB 70   (9 FOLDS)

L = 15.660	F0 = 4.10		
B = 8.540	F1 = -0.41		
t = 0.100	F2 = -0.94	M2 =	-1.12
d = 1.000	F3 = -0.48	M3 =	-5.53
	F4 = -1.74	M4 =	-4.97
	F5 = -1.04	M5 =	-1.42
	F6 = 11.56		

TABLE CUDEB 71   (9 FOLDS)

L = 16.130	F0 = 3.81		
B = 7.380	F1 = -0.17		
t = 0.100	F2 = -0.93	M2 =	-1.58
d = 1.000	F3 = -1.08	M3 =	-5.60
	F4 = -1.98	M4 =	-4.41
	F5 = -0.59	M5 =	-1.06
	F6 = 10.74		

TABLE CUDEB 72   (9 FOLDS)

L = 16.130	F0 = 3.92		
B = 7.600	F1 = -0.23		
t = 0.100	F2 = -0.93	M2 =	-1.52
d = 1.000	F3 = -0.97	M3 =	-5.68
	F4 = -1.95	M4 =	-4.57
	F5 = -0.67	M5 =	-1.13
	F6 = 11.00		

TABLE CUDEB 73   (9 FOLDS)

L = 16.130	F0 =	4.02		
B = 7.830	F1 =	-0.29		
t = 0.100	F2 =	-0.93	M2 =	-1.46
d = 1.000	F3 =	-0.86	M3 =	-5.74
	F4 =	-1.93	M4 =	-4.73
	F5 =	-0.76	M5 =	-1.20
	F6 =	11.28		

TABLE CUDEB 74   (9 FOLDS)

L = 16.130	F0 =	4.13		
B = 8.060	F1 =	-0.34		
t = 0.100	F2 =	-0.93	M2 =	-1.39
d = 1.000	F3 =	-0.75	M3 =	-5.80
	F4 =	-1.91	M4 =	-4.89
	F5 =	-0.85	M5 =	-1.27
	F6 =	11.55		

TABLE CUDEB 75   (9 FOLDS)

L = 16.130	F0 =	4.23		
B = 8.300	F1 =	-0.39		
t = 0.100	F2 =	-0.95	M2 =	-1.33
d = 1.000	F3 =	-0.65	M3 =	-5.84
	F4 =	-1.89	M4 =	-5.06
	F5 =	-0.94	M5 =	-1.34
	F6 =	11.84		

TABLE CUDEB 76   (9 FOLDS)

L = 16.130	F0 =	4.32		
B = 8.550	F1 =	-0.43		
t = 0.100	F2 =	-0.96	M2 =	-1.26
d = 1.000	F3 =	-0.54	M3 =	-5.87
	F4 =	-1.86	M4 =	-5.22
	F5 =	-1.03	M5 =	-1.43
	F6 =	12.14		

TABLE CUDEB 77   (9 FOLDS)

L = 16.130	F0 = 4.42	
B = 8.810	F1 = -0.46	
t = 0.100	F2 = -0.99	M2 = -1.21
d = 1.000	F3 = -0.43	M3 = -5.89
	F4 = -1.83	M4 = -5.39
	F5 = -1.12	M5 = -1.51
	F6 = 12.46	

TABLE CUDEB 78   (9 FOLDS)

L = 16.610	F0 = 4.09	
B = 7.600	F1 = -0.21	
t = 0.100	F2 = -0.96	M2 = -1.71
d = 1.000	F3 = -1.08	M3 = -6.01
	F4 = -2.08	M4 = -4.79
	F5 = -0.63	M5 = -1.13
	F6 = 11.52	

TABLE CUDEB 79   (9 FOLDS)

L = 16.610	F0 = 4.21	
B = 7.830	F1 = -0.28	
t = 0.100	F2 = -0.96	M2 = -1.64
d = 1.000	F3 = -0.96	M3 = -6.09
	F4 = -2.06	M4 = -4.97
	F5 = -0.72	M5 = -1.20
	F6 = 11.81	

TABLE CUDEB 80   (9 FOLDS)

L = 16.610	F0 = 4.32	
B = 8.060	F1 = -0.34	
t = 0.100	F2 = -0.96	M2 = -1.58
d = 1.000	F3 = -0.85	M3 = -6.15
	F4 = -2.04	M4 = -5.13
	F5 = -0.81	M5 = -1.27
	F6 = 12.10	

TABLE CUDEB 81   (9 FOLDS)

L = 16.610	F0 = 4.43		
B = 8.300	F1 = -0.39		
t = 0.100	F2 = -0.97	M2 = -1.51	
d = 1.000	F3 = -0.73	M3 = -6.19	
	F4 = -2.02	M4 = -5.31	
	F5 = -0.91	M5 = -1.34	
	F6 = 12.41		

TABLE CUDEB 82   (9 FOLDS)

L = 16.610	F0 = 4.54		
B = 8.550	F1 = -0.44		
t = 0.100	F2 = -0.99	M2 = -1.43	
d = 1.000	F3 = -0.62	M3 = -6.23	
	F4 = -1.99	M4 = -5.48	
	F5 = -1.00	M5 = -1.43	
	F6 = 12.73		

TABLE CUDEB 83   (9 FOLDS)

L = 16.610	F0 = 4.65		
B = 8.810	F1 = -0.48		
t = 0.100	F2 = -1.01	M2 = -1.37	
d = 1.000	F3 = -0.50	M3 = -6.25	
	F4 = -1.96	M4 = -5.66	
	F5 = -1.10	M5 = -1.51	
	F6 = 13.06		

TABLE CUDEB 84   (9 FOLDS)

L = 16.610	F0 = 4.75		
B = 9.070	F1 = -0.51		
t = 0.100	F2 = -1.04	M2 = -1.31	
d = 1.000	F3 = -0.39	M3 = -6.27	
	F4 = -1.94	M4 = -5.84	
	F5 = -1.20	M5 = -1.60	
	F6 = 13.39		

TABLE CUDEB 85   (9 FOLDS)

L =	17.110	F0 =	4.39			
B =	7.830	F1 =	−0.25			
t =	0.100	F2 =	−1.00	M2 =	−1.85	
d =	1.000	F3 =	−1.07	M3 =	−6.46	
		F4 =	−2.20	M4 =	−5.21	
		F5 =	−0.67	M5 =	−1.20	
		F6 =	12.37			

TABLE CUDEB 86   (9 FOLDS)

L =	17.110	F0 =	4.52			
B =	8.060	F1 =	−0.32			
t =	0.100	F2 =	−1.00	M2 =	−1.78	
d =	1.000	F3 =	−0.95	M3 =	−6.53	
		F4 =	−2.18	M4 =	−5.39	
		F5 =	−0.77	M5 =	−1.27	
		F6 =	12.68			

TABLE CUDEB 87   (9 FOLDS)

L =	17.110	F0 =	4.64			
B =	8.300	F1 =	−0.38			
t =	0.100	F2 =	−1.00	M2 =	−1.71	
d =	1.000	F3 =	−0.83	M3 =	−6.58	
		F4 =	−2.15	M4 =	−5.57	
		F5 =	−0.87	M5 =	−1.34	
		F6 =	13.00			

TABLE CUDEB 88   (9 FOLDS)

L =	17.110	F0 =	4.77			
B =	8.550	F1 =	−0.44			
t =	0.100	F2 =	−1.01	M2 =	−1.63	
d =	1.000	F3 =	−0.70	M3 =	−6.62	
		F4 =	−2.13	M4 =	−5.76	
		F5 =	−0.97	M5 =	−1.43	
		F6 =	13.34			

TABLE CUDEB 89 (9 FOLDS)
```
L =  17.110    F0 =    4.89
B =   8.810    F1 =   -0.49
t =   0.100    F2 =   -1.03    M2 =  -1.55
d =   1.000    F3 =   -0.58    M3 =  -6.65
               F4 =   -2.10    M4 =  -5.95
               F5 =   -1.07    M5 =  -1.51
               F6 =   13.69
```

TABLE CUDEB 90 (9 FOLDS)
```
L =  17.110    F0 =    5.00
B =   9.070    F1 =   -0.53
t =   0.100    F2 =   -1.06    M2 =  -1.48
d =   1.000    F3 =   -0.46    M3 =  -6.66
               F4 =   -2.08    M4 =  -6.14
               F5 =   -1.17    M5 =  -1.60
               F6 =   14.04
```

TABLE CUDEB 91 (9 FOLDS)
```
L =  17.110    F0 =    5.11
B =   9.340    F1 =   -0.56
t =   0.100    F2 =   -1.09    M2 =  -1.42
d =   1.000    F3 =   -0.34    M3 =  -6.67
               F4 =   -2.05    M4 =  -6.33
               F5 =   -1.28    M5 =  -1.70
               F6 =   14.41
```

TABLE CUDEB 92 (9 FOLDS)
```
L =  17.620    F0 =    4.71
B =   8.060    F1 =   -0.29
t =   0.100    F2 =   -1.03    M2 =  -2.00
d =   1.000    F3 =   -1.06    M3 =  -6.92
               F4 =   -2.32    M4 =  -5.65
               F5 =   -0.71    M5 =  -1.27
               F6 =   13.27
```

TABLE CUDEB 93   (9 FOLDS)

L = 17.620	F0 = 4.85		
B = 8.300	F1 = -0.36		
t = 0.100	F2 = -1.03	M2 = -1.92	
d = 1.000	F3 = -0.93	M3 = -6.99	
	F4 = -2.30	M4 = -5.85	
	F5 = -0.82	M5 = -1.34	
	F6 = 13.61		

TABLE CUDEB 94   (9 FOLDS)

L = 17.620	F0 = 4.99		
B = 8.550	F1 = -0.43		
t = 0.100	F2 = -1.04	M2 = -1.84	
d = 1.000	F3 = -0.80	M3 = -7.03	
	F4 = -2.28	M4 = -6.04	
	F5 = -0.92	M5 = -1.43	
	F6 = 13.96		

TABLE CUDEB 95   (9 FOLDS)

L = 17.620	F0 = 5.12		
B = 8.810	F1 = -0.49		
t = 0.100	F2 = -1.06	M2 = -1.76	
d = 1.000	F3 = -0.66	M3 = -7.07	
	F4 = -2.25	M4 = -6.25	
	F5 = -1.03	M5 = -1.51	
	F6 = 14.33		

TABLE CUDEB 96   (9 FOLDS)

L = 17.620	F0 = 5.25		
B = 9.070	F1 = -0.54		
t = 0.100	F2 = -1.08	M2 = -1.68	
d = 1.000	F3 = -0.54	M3 = -7.09	
	F4 = -2.23	M4 = -6.45	
	F5 = -1.14	M5 = -1.60	
	F6 = 14.70		

TABLE CUDEB 97 (9 FOLDS)

L = 17.620	F0 =	5.37		
B = 9.340	F1 =	-0.59		
t = 0.100	F2 =	-1.11	M2 =	-1.60
d = 1.000	F3 =	-0.41	M3 =	-7.09
	F4 =	-2.20	M4 =	-6.65
	F5 =	-1.25	M5 =	-1.70
	F6 =	15.09		

TABLE CUDEB 98 (9 FOLDS)

L = 17.620	F0 =	5.49		
B = 9.620	F1 =	-0.62		
t = 0.100	F2 =	-1.15	M2 =	-1.54
d = 1.000	F3 =	-0.29	M3 =	-7.09
	F4 =	-2.17	M4 =	-6.86
	F5 =	-1.36	M5 =	-1.80
	F6 =	15.49		

TABLE CUDEB 99 (9 FOLDS)

L = 18.150	F0 =	5.06		
B = 8.300	F1 =	-0.33		
t = 0.100	F2 =	-1.07	M2 =	-2.16
d = 1.000	F3 =	-1.05	M3 =	-7.42
	F4 =	-2.45	M4 =	-6.13
	F5 =	-0.76	M5 =	-1.34
	F6 =	14.24		

TABLE CUDEB 100 (9 FOLDS)

L = 18.150	F0 =	5.21		
B = 8.550	F1 =	-0.41		
t = 0.100	F2 =	-1.08	M2 =	-2.08
d = 1.000	F3 =	-0.91	M3 =	-7.48
	F4 =	-2.43	M4 =	-6.34
	F5 =	-0.87	M5 =	-1.43
	F6 =	14.61		

TABLE CUDEB 101   (9 FOLDS)

L = 18.150	F0 =	5.36		
B = 8.810	F1 =	-0.49		
t = 0.100	F2 =	-1.09	M2 =	-1.99
d = 1.000	F3 =	-0.76	M3 =	-7.52
	F4 =	-2.41	M4 =	-6.56
	F5 =	-0.99	M5 =	-1.51
	F6 =	15.01		

TABLE CUDEB 102   (9 FOLDS)

L = 18.150	F0 =	5.51		
B = 9.070	F1 =	-0.55		
t = 0.100	F2 =	-1.11	M2 =	-1.90
d = 1.000	F3 =	-0.63	M3 =	-7.55
	F4 =	-2.38	M4 =	-6.77
	F5 =	-1.10	M5 =	-1.60
	F6 =	15.40		

TABLE CUDEB 103   (9 FOLDS)

L = 18.150	F0 =	5.64		
B = 9.340	F1 =	-0.60		
t = 0.100	F2 =	-1.14	M2 =	-1.81
d = 1.000	F3 =	-0.49	M3 =	-7.56
	F4 =	-2.35	M4 =	-6.99
	F5 =	-1.21	M5 =	-1.70
	F6 =	15.81		

TABLE CUDEB 104   (9 FOLDS)

L = 18.150	F0 =	5.78		
B = 9.620	F1 =	-0.65		
t = 0.100	F2 =	-1.17	M2 =	-1.73
d = 1.000	F3 =	-0.36	M3 =	-7.55
	F4 =	-2.33	M4 =	-7.21
	F5 =	-1.33	M5 =	-1.80
	F6 =	16.23		

TABLE CUDEB 105   (9 FOLDS)

L = 18.150	F0 = 5.90		
B = 9.910	F1 = -0.68		
t = 0.100	F2 = -1.21	M2 = -1.67
d = 1.000	F3 = -0.23	M3 = -7.54
	F4 = -2.29	M4 = -7.43
	F5 = -1.45	M5 = -1.91
	F6 = 16.67		

TABLE CUDEB 106   (9 FOLDS)

L = 18.690	F0 = 5.44		
B = 8.550	F1 = -0.39		
t = 0.100	F2 = -1.11	M2 = -2.33
d = 1.000	F3 = -1.03	M3 = -7.95
	F4 = -2.59	M4 = -6.65
	F5 = -0.80	M5 = -1.43
	F6 = 15.28		

TABLE CUDEB 107   (9 FOLDS)

L = 18.690	F0 = 5.60		
B = 8.810	F1 = -0.47		
t = 0.100	F2 = -1.12	M2 = -2.24
d = 1.000	F3 = -0.87	M3 = -8.00
	F4 = -2.57	M4 = -6.88
	F5 = -0.93	M5 = -1.51
	F6 = 15.69		

TABLE CUDEB 108   (9 FOLDS)

L = 18.690	F0 = 5.76		
B = 9.070	F1 = -0.55		
t = 0.100	F2 = -1.14	M2 = -2.14
d = 1.000	F3 = -0.73	M3 = -8.03
	F4 = -2.54	M4 = -7.10
	F5 = -1.05	M5 = -1.60
	F6 = 16.11		

TABLE CUDEB 109　(9 FOLDS)

L = 18.690	F0 = 5.92	
B = 9.340	F1 = -0.61	
t = 0.100	F2 = -1.16	M2 = -2.05
d = 1.000	F3 = -0.58	M3 = -8.05
	F4 = -2.52	M4 = -7.33
	F5 = -1.17	M5 = -1.70
	F6 = 16.54	

TABLE CUDEB 110　(9 FOLDS)

L = 18.690	F0 = 6.07	
B = 9.620	F1 = -0.67	
t = 0.100	F2 = -1.19	M2 = -1.96
d = 1.000	F3 = -0.44	M3 = -8.05
	F4 = -2.49	M4 = -7.57
	F5 = -1.29	M5 = -1.80
	F6 = 16.99	

TABLE CUDEB 111　(9 FOLDS)

L = 18.690	F0 = 6.21	
B = 9.910	F1 = -0.71	
t = 0.100	F2 = -1.23	M2 = -1.87
d = 1.000	F3 = -0.29	M3 = -8.03
	F4 = -2.46	M4 = -7.81
	F5 = -1.42	M5 = -1.91
	F6 = 17.45	

TABLE CUDEB 112　(9 FOLDS)

L = 18.690	F0 = 6.35	
B = 10.210	F1 = -0.74	
t = 0.100	F2 = -1.28	M2 = -1.81
d = 1.000	F3 = -0.15	M3 = -8.00
	F4 = -2.43	M4 = -8.05
	F5 = -1.55	M5 = -2.03
	F6 = 17.94	

TABLE CUDEB 113   (9 FOLDS)

L = 19.250	F0 =	5.84		
B = 8.810	F1 =	-0.44		
t = 0.100	F2 =	-1.15	M2 =	-2.51
d = 1.000	F3 =	-1.00	M3 =	-8.51
	F4 =	-2.74	M4 =	-7.22
	F5 =	-0.85	M5 =	-1.51
	F6 =	16.40		

TABLE CUDEB 114   (9 FOLDS)

L = 19.250	F0 =	6.02		
B = 9.070	F1 =	-0.53		
t = 0.100	F2 =	-1.17	M2 =	-2.41
d = 1.000	F3 =	-0.84	M3 =	-8.55
	F4 =	-2.71	M4 =	-7.46
	F5 =	-0.98	M5 =	-1.60
	F6 =	16.84		

TABLE CUDEB 115   (9 FOLDS)

L = 19.250	F0 =	6.19		
B = 9.340	F1 =	-0.61		
t = 0.100	F2 =	-1.19	M2 =	-2.31
d = 1.000	F3 =	-0.68	M3 =	-8.58
	F4 =	-2.69	M4 =	-7.70
	F5 =	-1.11	M5 =	-1.70
	F6 =	17.30		

TABLE CUDEB 116   (9 FOLDS)

L = 19.250	F0 =	6.36		
B = 9.620	F1 =	-0.68		
t = 0.100	F2 =	-1.22	M2 =	-2.21
d = 1.000	F3 =	-0.53	M3 =	-8.58
	F4 =	-2.67	M4 =	-7.95
	F5 =	-1.24	M5 =	-1.80
	F6 =	17.77		

TABLE CUDEB 117   (9 FOLDS)

L = 19.250	F0 = 6.52		
B = 9.910	F1 = -0.73		
t = 0.100	F2 = -1.25	M2 = -2.11
d = 1.000	F3 = -0.37	M3 = -8.57
	F4 = -2.64	M4 = -8.20
	F5 = -1.37	M5 = -1.91
	F6 = 18.26		

TABLE CUDEB 118   (9 FOLDS)

L = 19.250	F0 = 6.68		
B = 10.210	F1 = -0.78		
t = 0.100	F2 = -1.30	M2 = -2.03
d = 1.000	F3 = -0.22	M3 = -8.54
	F4 = -2.61	M4 = -8.46
	F5 = -1.51	M5 = -2.03
	F6 = 18.77		

TABLE CUDEB 119   (9 FOLDS)

L = 19.250	F0 = 6.83		
B = 10.520	F1 = -0.81		
t = 0.100	F2 = -1.35	M2 = -1.96
d = 1.000	F3 = -0.07	M3 = -8.49
	F4 = -2.57	M4 = -8.72
	F5 = -1.65	M5 = -2.16
	F6 = 19.30		

TABLE CUDEB 120   (9 FOLDS)

L = 19.830	F0 = 6.27		
B = 9.070	F1 = -0.50		
t = 0.100	F2 = -1.20	M2 = -2.71
d = 1.000	F3 = -0.97	M3 = -9.11
	F4 = -2.90	M4 = -7.82
	F5 = -0.90	M5 = -1.60
	F6 = 17.60		

TABLE CUDEB 121   (9 FOLDS)
L = 19.830   F0 =   6.47
B = 9.340   F1 =   -0.59
t = 0.100   F2 =   -1.22   M2 =   -2.60
d = 1.000   F3 =   -0.80   M3 =   -9.14
              F4 =   -2.87   M4 =   -8.08
              F5 =   -1.04   M5 =   -1.70
              F6 =   18.08

TABLE CUDEB 122   (9 FOLDS)
L = 19.830   F0 =   6.66
B = 9.620   F1 =   -0.68
t = 0.100   F2 =   -1.24   M2 =   -2.49
d = 1.000   F3 =   -0.63   M3 =   -9.16
              F4 =   -2.85   M4 =   -8.34
              F5 =   -1.18   M5 =   -1.80
              F6 =   18.58

TABLE CUDEB 123   (9 FOLDS)
L = 19.830   F0 =   6.84
B = 9.910   F1 =   -0.75
t = 0.100   F2 =   -1.28   M2 =   -2.38
d = 1.000   F3 =   -0.46   M3 =   -9.15
              F4 =   -2.82   M4 =   -8.61
              F5 =   -1.32   M5 =   -1.91
              F6 =   19.10

TABLE CUDEB 124   (9 FOLDS)
L = 19.830   F0 =   7.02
B = 10.210   F1 =   -0.81
t = 0.100   F2 =   -1.32   M2 =   -2.28
d = 1.000   F3 =   -0.30   M3 =   -9.12
              F4 =   -2.80   M4 =   -8.89
              F5 =   -1.46   M5 =   -2.03
              F6 =   19.64

TABLE CUDEB 125   (9 FOLDS)

L = 19.830	F0 = 7.19	
B = 10.520	F1 = -0.85	
t = 0.100	F2 = -1.37	M2 = -2.19
d = 1.000	F3 = -0.14	M3 = -9.08
	F4 = -2.76	M4 = -9.17
	F5 = -1.60	M5 = -2.16
	F6 = 20.20	

TABLE CUDEB 126   (9 FOLDS)

L = 19.830	F0 = 7.35	
B = 10.840	F1 = -0.88	
t = 0.100	F2 = -1.44	M2 = -2.12
d = 1.000	F3 = 0.02	M3 = -9.01
	F4 = -2.73	M4 = -9.45
	F5 = -1.75	M5 = -2.29
	F6 = 20.78	

TABLE CUDEB 127   (9 FOLDS)

L = 20.420	F0 = 6.74	
B = 9.340	F1 = -0.56	
t = 0.100	F2 = -1.25	M2 = -2.92
d = 1.000	F3 = -0.93	M3 = -9.74
	F4 = -3.06	M4 = -8.47
	F5 = -0.95	M5 = -1.70
	F6 = 18.88	

TABLE CUDEB 128   (9 FOLDS)

L = 20.420	F0 = 6.95	
B = 9.620	F1 = -0.66	
t = 0.100	F2 = -1.27	M2 = -2.80
d = 1.000	F3 = -0.75	M3 = -9.76
	F4 = -3.04	M4 = -8.75
	F5 = -1.10	M5 = -1.80
	F6 = 19.41	

TABLE CUDEB 129 (9 FOLDS)

L = 20.420	F0 = 7.15			
B = 9.910	F1 = -0.75			
t = 0.100	F2 = -1.30	M2 = -2.68		
d = 1.000	F3 = -0.57	M3 = -9.76		
	F4 = -3.02	M4 = -9.04		
	F5 = -1.25	M5 = -1.91		
	F6 = 19.96			

TABLE CUDEB 130 (9 FOLDS)

L = 20.420	F0 = 7.35	
B = 10.210	F1 = -0.82	
t = 0.100	F2 = -1.35	M2 = -2.56
d = 1.000	F3 = -0.39	M3 = -9.74
	F4 = -2.99	M4 = -9.33
	F5 = -1.40	M5 = -2.03
	F6 = 20.53	

TABLE CUDEB 131 (9 FOLDS)

L = 20.420	F0 = 7.55	
B = 10.520	F1 = -0.88	
t = 0.100	F2 = -1.40	M2 = -2.45
d = 1.000	F3 = -0.22	M3 = -9.70
	F4 = -2.96	M4 = -9.62
	F5 = -1.55	M5 = -2.16
	F6 = 21.12	

TABLE CUDEB 132 (9 FOLDS)

L = 20.420	F0 = 7.73	
B = 10.840	F1 = -0.93	
t = 0.100	F2 = -1.45	M2 = -2.36
d = 1.000	F3 = -0.05	M3 = -9.63
	F4 = -2.93	M4 = -9.93
	F5 = -1.70	M5 = -2.29
	F6 = 21.73	

TABLE CUDEB 133   (9 FOLDS)

```
L =  20.420    F0 =    7.90
B =  11.170    F1 =   -0.95
t =   0.100    F2 =   -1.52    M2 =   -2.30
d =   1.000    F3 =    0.12    M3 =   -9.55
               F4 =   -2.89    M4 =  -10.24
               F5 =   -1.86    M5 =   -2.43
               F6 =   22.35
```

TABLE CUDEB 134   (9 FOLDS)

```
L =  21.030    F0 =    7.24
B =   9.620    F1 =   -0.63
t =   0.100    F2 =   -1.30    M2 =   -3.14
d =   1.000    F3 =   -0.88    M3 =  -10.41
               F4 =   -3.24    M4 =   -9.18
               F5 =   -1.00    M5 =   -1.80
               F6 =   20.26
```

TABLE CUDEB 135   (9 FOLDS)

```
L =  21.030    F0 =    7.47
B =   9.910    F1 =   -0.74
t =   0.100    F2 =   -1.33    M2 =   -3.01
d =   1.000    F3 =   -0.69    M3 =  -10.42
               F4 =   -3.22    M4 =   -9.48
               F5 =   -1.16    M5 =   -1.91
               F6 =   20.84
```

TABLE CUDEB 136   (9 FOLDS)

```
L =  21.030    F0 =    7.69
B =  10.210    F1 =   -0.83
t =   0.100    F2 =   -1.37    M2 =   -2.88
d =   1.000    F3 =   -0.50    M3 =  -10.41
               F4 =   -3.20    M4 =   -9.79
               F5 =   -1.32    M5 =   -2.03
               F6 =   21.44
```

TABLE CUDEB 137   (9 FOLDS)

L = 21.030	F0 = 7.91	
B = 10.520	F1 = -0.90	
t = 0.100	F2 = -1.42	M2 = -2.75
d = 1.000	F3 = -0.31	M3 = -10.37
	F4 = -3.17	M4 = -10.10
	F5 = -1.48	M5 = -2.16
	F6 = 22.06	

TABLE CUDEB 138   (9 FOLDS)

L = 21.030	F0 = 8.12	
B = 10.840	F1 = -0.96	
t = 0.100	F2 = -1.47	M2 = -2.64
d = 1.000	F3 = -0.12	M3 = -10.31
	F4 = -3.14	M4 = -10.42
	F5 = -1.64	M5 = -2.29
	F6 = 22.71	

TABLE CUDEB 139   (9 FOLDS)

L = 21.030	F0 = 8.31	
B = 11.170	F1 = -1.01	
t = 0.100	F2 = -1.54	M2 = -2.55
d = 1.000	F3 = 0.06	M3 = -10.22
	F4 = -3.10	M4 = -10.75
	F5 = -1.81	M5 = -2.43
	F6 = 23.37	

TABLE CUDEB 140   (9 FOLDS)

L = 21.030	F0 = 8.49	
B = 11.510	F1 = -1.03	
t = 0.100	F2 = -1.62	M2 = -2.49
d = 1.000	F3 = 0.23	M3 = -10.12
	F4 = -3.06	M4 = -11.09
	F5 = -1.97	M5 = -2.58
	F6 = 24.05	

Appendix 8. Design tables for folded plates types CPDEB[1]

```
TABLE CPDEB 1  (9 FOLDS)
L =  12.000    F0 =     0.47
B =   5.490    F1 =     1.04
t =   0.100    F2 =    -0.92    M2 =    -1.45
d =   0.200    F3 =    -1.11    M3 =    -3.67
               F4 =    -1.07    M4 =    -2.37
               F5 =    -0.19    M5 =    -0.59
               F6 =     5.08


TABLE CPDEB 2  (9 FOLDS)
L =  12.000    F0 =     0.43
B =   5.650    F1 =     1.07
t =   0.100    F2 =    -0.93    M2 =    -1.49
d =   0.200    F3 =    -1.05    M3 =    -3.81
               F4 =    -1.04    M4 =    -2.46
               F5 =    -0.24    M5 =    -0.62
               F6 =     5.19


TABLE CPDEB 3  (9 FOLDS)
L =  12.000    F0 =     0.39
B =   5.820    F1 =     1.09
t =   0.100    F2 =    -0.95    M2 =    -1.54
d =   0.200    F3 =    -0.98    M3 =    -3.95
               F4 =    -1.01    M4 =    -2.56
               F5 =    -0.30    M5 =    -0.66
               F6 =     5.31


TABLE CPDEB 4  (9 FOLDS)
L =  12.000    F0 =     0.36
B =   5.990    F1 =     1.12
t =   0.100    F2 =    -0.97    M2 =    -1.60
d =   0.200    F3 =    -0.91    M3 =    -4.08
               F4 =    -0.98    M4 =    -2.65
               F5 =    -0.36    M5 =    -0.70
               F6 =     5.44
```

[1] See also Figs A7 and A8, p. 192

TABLE CPDEB 5   (9 FOLDS)

L = 12.000	F0 =	0.31		
B = 6.170	F1 =	1.15		
t = 0.100	F2 =	-1.00	M2 =	-1.65
d = 0.200	F3 =	-0.84	M3 =	-4.21
	F4 =	-0.95	M4 =	-2.73
	F5 =	-0.43	M5 =	-0.74
	F6 =	5.58		

TABLE CPDEB 6   (9 FOLDS)

L = 12.000	F0 =	0.27		
B = 6.360	F1 =	1.19		
t = 0.100	F2 =	-1.03	M2 =	-1.71
d = 0.200	F3 =	-0.76	M3 =	-4.33
	F4 =	-0.92	M4 =	-2.80
	F5 =	-0.50	M5 =	-0.79
	F6 =	5.73		

TABLE CPDEB 7   (9 FOLDS)

L = 12.000	F0 =	0.23		
B = 6.550	F1 =	1.22		
t = 0.100	F2 =	-1.07	M2 =	-1.77
d = 0.200	F3 =	-0.69	M3 =	-4.43
	F4 =	-0.88	M4 =	-2.86
	F5 =	-0.57	M5 =	-0.84
	F6 =	5.88		

TABLE CPDEB 8   (9 FOLDS)

L = 12.360	F0 =	0.50		
B = 5.650	F1 =	1.09		
t = 0.100	F2 =	-0.94	M2 =	-1.51
d = 0.200	F3 =	-1.13	M3 =	-3.89
	F4 =	-1.13	M4 =	-2.54
	F5 =	-0.21	M5 =	-0.62
	F6 =	5.46		

TABLE CPDEB 9   (9 FOLDS)

L = 12.360	F0 =	0.46		
B = 5.820	F1 =	1.11		
t = 0.100	F2 =	-0.95	M2 =	-1.56
d = 0.200	F3 =	-1.06	M3 =	-4.04
	F4 =	-1.10	M4 =	-2.65
	F5 =	-0.27	M5 =	-0.66
	F6 =	5.58		

TABLE CPDEB 10   (9 FOLDS)

L = 12.360	F0 =	0.42		
B = 5.990	F1 =	1.14		
t = 0.100	F2 =	-0.97	M2 =	-1.61
d = 0.200	F3 =	-0.99	M3 =	-4.19
	F4 =	-1.07	M4 =	-2.75
	F5 =	-0.33	M5 =	-0.70
	F6 =	5.71		

TABLE CPDEB 11   (9 FOLDS)

L = 12.360	F0 =	0.38		
B = 6.170	F1 =	1.17		
t = 0.100	F2 =	-1.00	M2 =	-1.67
d = 0.200	F3 =	-0.92	M3 =	-4.33
	F4 =	-1.03	M4 =	-2.84
	F5 =	-0.40	M5 =	-0.74
	F6 =	5.85		

TABLE CPDEB 12   (9 FOLDS)

L = 12.360	F0 =	0.33		
B = 6.360	F1 =	1.21		
t = 0.100	F2 =	-1.03	M2 =	-1.73
d = 0.200	F3 =	-0.84	M3 =	-4.46
	F4 =	-1.00	M4 =	-2.93
	F5 =	-0.47	M5 =	-0.79
	F6 =	6.01		

TABLE CPDEB 13   (9 FOLDS)

L = 12.360	F0 = 0.29
B = 6.550	F1 = 1.24
t = 0.100	F2 = -1.07 M2 = -1.79
d = 0.200	F3 = -0.77 M3 = -4.58
	F4 = -0.97 M4 = -3.01
	F5 = -0.55 M5 = -0.84
	F6 = 6.16

TABLE CPDEB 14   (9 FOLDS)

L = 12.360	F0 = 0.25
B = 6.750	F1 = 1.28
t = 0.100	F2 = -1.11 M2 = -1.86
d = 0.200	F3 = -0.69 M3 = -4.69
	F4 = -0.93 M4 = -3.07
	F5 = -0.62 M5 = -0.89
	F6 = 6.33

TABLE CPDEB 15   (9 FOLDS)

L = 12.730	F0 = 0.53
B = 5.820	F1 = 1.14
t = 0.100	F2 = -0.96 M2 = -1.57
d = 0.200	F3 = -1.16 M3 = -4.13
	F4 = -1.19 M4 = -2.73
	F5 = -0.23 M5 = -0.66
	F6 = 5.87

TABLE CPDEB 16   (9 FOLDS)

L = 12.730	F0 = 0.49
B = 5.990	F1 = 1.17
t = 0.100	F2 = -0.98 M2 = -1.62
d = 0.200	F3 = -1.08 M3 = -4.28
	F4 = -1.16 M4 = -2.85
	F5 = -0.29 M5 = -0.70
	F6 = 6.00

TABLE CPDEB 17   (9 FOLDS)

L = 12.730	F0 =	0.44		
B = 6.170	F1 =	1.20		
t = 0.100	F2 =	-1.00	M2 =	-1.68
d = 0.200	F3 =	-1.01	M3 =	-4.44
	F4 =	-1.13	M4 =	-2.96
	F5 =	-0.36	M5 =	-0.74
	F6 =	6.14		

TABLE CPDEB 18   (9 FOLDS)

L = 12.730	F0 =	0.40		
B = 6.360	F1 =	1.23		
t = 0.100	F2 =	-1.03	M2 =	-1.74
d = 0.200	F3 =	-0.93	M3 =	-4.59
	F4 =	-1.09	M4 =	-3.06
	F5 =	-0.44	M5 =	-0.79
	F6 =	6.30		

TABLE CPDEB 19   (9 FOLDS)

L = 12.730	F0 =	0.36		
B = 6.550	F1 =	1.26		
t = 0.100	F2 =	-1.07	M2 =	-1.80
d = 0.200	F3 =	-0.85	M3 =	-4.72
	F4 =	-1.05	M4 =	-3.15
	F5 =	-0.51	M5 =	-0.84
	F6 =	6.46		

TABLE CPDEB 20   (9 FOLDS)

L = 12.730	F0 =	0.31		
B = 6.750	F1 =	1.30		
t = 0.100	F2 =	-1.11	M2 =	-1.87
d = 0.200	F3 =	-0.76	M3 =	-4.85
	F4 =	-1.02	M4 =	-3.23
	F5 =	-0.59	M5 =	-0.89
	F6 =	6.64		

TABLE CPDEB 21   (9 FOLDS)

```
L =  12.730    F0 =    0.27
B =   6.950    F1 =    1.34
t =   0.100    F2 =   -1.16    M2 =   -1.94
d =   0.200    F3 =   -0.68    M3 =   -4.95
               F4 =   -0.98    M4 =   -3.30
               F5 =   -0.68    M5 =   -0.94
               F6 =    6.81
```

TABLE CPDEB 22   (9 FOLDS)

```
L =  13.110    F0 =    0.56
B =   6.000    F1 =    1.19
t =   0.100    F2 =   -0.98    M2 =   -1.64
d =   0.200    F3 =   -1.17    M3 =   -4.39
               F4 =   -1.26    M4 =   -2.95
               F5 =   -0.26    M5 =   -0.70
               F6 =    6.31
```

TABLE CPDEB 23   (9 FOLDS)

```
L =  13.110    F0 =    0.51
B =   6.180    F1 =    1.22
t =   0.100    F2 =   -1.00    M2 =   -1.70
d =   0.200    F3 =   -1.10    M3 =   -4.55
               F4 =   -1.22    M4 =   -3.07
               F5 =   -0.33    M5 =   -0.74
               F6 =    6.45
```

TABLE CPDEB 24   (9 FOLDS)

```
L =  13.110    F0 =    0.47
B =   6.370    F1 =    1.25
t =   0.100    F2 =   -1.03    M2 =   -1.76
d =   0.200    F3 =   -1.01    M3 =   -4.72
               F4 =   -1.18    M4 =   -3.19
               F5 =   -0.40    M5 =   -0.79
               F6 =    6.62
```

TABLE CPDEB 25   (9 FOLDS)

L = 13.110	F0 = 0.42		
B = 6.560	F1 = 1.29		
t = 0.100	F2 = -1.07	M2 = -1.82
d = 0.200	F3 = -0.93	M3 = -4.87
	F4 = -1.15	M4 = -3.29
	F5 = -0.48	M5 = -0.84
	F6 = 6.78		

TABLE CPDEB 26   (9 FOLDS)

L = 13.110	F0 = 0.37		
B = 6.760	F1 = 1.33		
t = 0.100	F2 = -1.11	M2 = -1.89
d = 0.200	F3 = -0.84	M3 = -5.01
	F4 = -1.11	M4 = -3.39
	F5 = -0.56	M5 = -0.89
	F6 = 6.96		

TABLE CPDEB 27   (9 FOLDS)

L = 13.110	F0 = 0.33		
B = 6.960	F1 = 1.37		
t = 0.100	F2 = -1.16	M2 = -1.96
d = 0.200	F3 = -0.76	M3 = -5.13
	F4 = -1.07	M4 = -3.48
	F5 = -0.65	M5 = -0.94
	F6 = 7.15		

TABLE CPDEB 28   (9 FOLDS)

L = 13.110	F0 = 0.28		
B = 7.170	F1 = 1.41		
t = 0.100	F2 = -1.21	M2 = -2.04
d = 0.200	F3 = -0.67	M3 = -5.24
	F4 = -1.04	M4 = -3.55
	F5 = -0.74	M5 = -1.00
	F6 = 7.35		

TABLE CPDEB 29   (9 FOLDS)

L = 13.500	F0 = 0.59	
B = 6.180	F1 = 1.25	
t = 0.100	F2 = -1.00	M2 = -1.71
d = 0.200	F3 = -1.19	M3 = -4.65
	F4 = -1.32	M4 = -3.17
	F5 = -0.29	M5 = -0.74
	F6 = 6.78	

TABLE CPDEB 30   (9 FOLDS)

L = 13.500	F0 = 0.54	
B = 6.370	F1 = 1.28	
t = 0.100	F2 = -1.03	M2 = -1.77
d = 0.200	F3 = -1.11	M3 = -4.83
	F4 = -1.29	M4 = -3.31
	F5 = -0.36	M5 = -0.79
	F6 = 6.94	

TABLE CPDEB 31   (9 FOLDS)

L = 13.500	F0 = 0.49	
B = 6.560	F1 = 1.31	
t = 0.100	F2 = -1.06	M2 = -1.83
d = 0.200	F3 = -1.02	M3 = -5.00
	F4 = -1.25	M4 = -3.43
	F5 = -0.44	M5 = -0.84
	F6 = 7.11	

TABLE CPDEB 32   (9 FOLDS)

L = 13.500	F0 = 0.44	
B = 6.760	F1 = 1.35	
t = 0.100	F2 = -1.11	M2 = -1.90
d = 0.200	F3 = -0.93	M3 = -5.16
	F4 = -1.21	M4 = -3.54
	F5 = -0.53	M5 = -0.89
	F6 = 7.29	

TABLE CPDEB 33   (9 FOLDS)

L = 13.500	F0 =	0.40		
B = 6.960	F1 =	1.39		
t = 0.100	F2 =	−1.15	M2 =	−1.98
d = 0.200	F3 =	−0.85	M3 =	−5.30
	F4 =	−1.17	M4 =	−3.64
	F5 =	−0.61	M5 =	−0.94
	F6 =	7.49		

TABLE CPDEB 34   (9 FOLDS)

L = 13.500	F0 =	0.35		
B = 7.170	F1 =	1.43		
t = 0.100	F2 =	−1.20	M2 =	−2.05
d = 0.200	F3 =	−0.75	M3 =	−5.43
	F4 =	−1.13	M4 =	−3.74
	F5 =	−0.70	M5 =	−1.00
	F6 =	7.69		

TABLE CPDEB 35   (9 FOLDS)

L = 13.500	F0 =	0.30		
B = 7.390	F1 =	1.48		
t = 0.100	F2 =	−1.26	M2 =	−2.14
d = 0.200	F3 =	−0.66	M3 =	−5.54
	F4 =	−1.09	M4 =	−3.82
	F5 =	−0.80	M5 =	−1.06
	F6 =	7.91		

TABLE CPDEB 36   (9 FOLDS)

L = 13.910	F0 =	0.62		
B = 6.360	F1 =	1.30		
t = 0.100	F2 =	−1.03	M2 =	−1.78
d = 0.200	F3 =	−1.22	M3 =	−4.94
	F4 =	−1.40	M4 =	−3.41
	F5 =	−0.31	M5 =	−0.79
	F6 =	7.28		

TABLE CPDEB 37   (9 FOLDS)

L =	13.910	F0 =	0.57		
B =	6.550	F1 =	1.34		
t =	0.100	F2 =	-1.06	M2 =	-1.84
d =	0.200	F3 =	-1.13	M3 =	-5.12
		F4 =	-1.36	M4 =	-3.55
		F5 =	-0.39	M5 =	-0.84
		F6 =	7.45		

TABLE CPDEB 38   (9 FOLDS)

L =	13.910	F0 =	0.52		
B =	6.750	F1 =	1.37		
t =	0.100	F2 =	-1.10	M2 =	-1.91
d =	0.200	F3 =	-1.04	M3 =	-5.30
		F4 =	-1.32	M4 =	-3.69
		F5 =	-0.48	M5 =	-0.89
		F6 =	7.64		

TABLE CPDEB 39   (9 FOLDS)

L =	13.910	F0 =	0.47		
B =	6.950	F1 =	1.41		
t =	0.100	F2 =	-1.14	M2 =	-1.99
d =	0.200	F3 =	-0.95	M3 =	-5.46
		F4 =	-1.28	M4 =	-3.81
		F5 =	-0.56	M5 =	-0.94
		F6 =	7.84		

TABLE CPDEB 40   (9 FOLDS)

L =	13.910	F0 =	0.42		
B =	7.160	F1 =	1.46		
t =	0.100	F2 =	-1.19	M2 =	-2.06
d =	0.200	F3 =	-0.85	M3 =	-5.61
		F4 =	-1.24	M4 =	-3.92
		F5 =	-0.66	M5 =	-1.00
		F6 =	8.05		

TABLE CPDEB 41   (9 FOLDS)

L = 13.910	F0 = 0.38	
B = 7.370	F1 = 1.50	
t = 0.100	F2 = -1.25	M2 = -2.14
d = 0.200	F3 = -0.75	M3 = -5.74
	F4 = -1.20	M4 = -4.01
	F5 = -0.75	M5 = -1.06
	F6 = 8.27	

TABLE CPDEB 42   (9 FOLDS)

L = 13.910	F0 = 0.33	
B = 7.590	F1 = 1.54	
t = 0.100	F2 = -1.31	M2 = -2.23
d = 0.200	F3 = -0.65	M3 = -5.85
	F4 = -1.16	M4 = -4.10
	F5 = -0.85	M5 = -1.12
	F6 = 8.50	

TABLE CPDEB 43   (9 FOLDS)

L = 14.330	F0 = 0.65	
B = 6.560	F1 = 1.36	
t = 0.100	F2 = -1.05	M2 = -1.86
d = 0.200	F3 = -1.23	M3 = -5.25
	F4 = -1.48	M4 = -3.69
	F5 = -0.34	M5 = -0.84
	F6 = 7.83	

TABLE CPDEB 44   (9 FOLDS)

L = 14.330	F0 = 0.60	
B = 6.760	F1 = 1.40	
t = 0.100	F2 = -1.09	M2 = -1.93
d = 0.200	F3 = -1.14	M3 = -5.45
	F4 = -1.44	M4 = -3.84
	F5 = -0.43	M5 = -0.89
	F6 = 8.03	

TABLE CPDEB 45   (9 FOLDS)

L = 14.330	F0 =	0.55		
B = 6.960	F1 =	1.44		
t = 0.100	F2 =	-1.13	M2 =	-2.00
d = 0.200	F3 =	-1.04	M3 =	-5.63
	F4 =	-1.39	M4 =	-3.98
	F5 =	-0.52	M5 =	-0.94
	F6 =	8.23		

TABLE CPDEB 46   (9 FOLDS)

L = 14.330	F0 =	0.50		
B = 7.170	F1 =	1.48		
t = 0.100	F2 =	-1.19	M2 =	-2.08
d = 0.200	F3 =	-0.94	M3 =	-5.80
	F4 =	-1.35	M4 =	-4.11
	F5 =	-0.62	M5 =	-1.00
	F6 =	8.44		

TABLE CPDEB 47   (9 FOLDS)

L = 14.330	F0 =	0.45		
B = 7.390	F1 =	1.53		
t = 0.100	F2 =	-1.24	M2 =	-2.16
d = 0.200	F3 =	-0.84	M3 =	-5.95
	F4 =	-1.31	M4 =	-4.22
	F5 =	-0.72	M5 =	-1.06
	F6 =	8.68		

TABLE CPDEB 48   (9 FOLDS)

L = 14.330	F0 =	0.39		
B = 7.610	F1 =	1.58		
t = 0.100	F2 =	-1.31	M2 =	-2.25
d = 0.200	F3 =	-0.73	M3 =	-6.08
	F4 =	-1.27	M4 =	-4.33
	F5 =	-0.82	M5 =	-1.13
	F6 =	8.92		

TABLE CPDEB 49   (9 FOLDS)

```
L = 14.330    F0 =    0.35
B =  7.840    F1 =    1.62
t =  0.100    F2 =   -1.37    M2 =   -2.35
d =  0.200    F3 =   -0.63    M3 =   -6.20
              F4 =   -1.23    M4 =   -4.42
              F5 =   -0.93    M5 =   -1.20
              F6 =    9.18
```

TABLE CPDEB 50   (9 FOLDS)

```
L = 14.760    F0 =    0.69
B =  6.750    F1 =    1.43
t =  0.100    F2 =   -1.08    M2 =   -1.94
d =  0.200    F3 =   -1.25    M3 =   -5.57
              F4 =   -1.56    M4 =   -3.97
              F5 =   -0.37    M5 =   -0.89
              F6 =    8.41
```

TABLE CPDEB 51   (9 FOLDS)

```
L = 14.760    F0 =    0.64
B =  6.950    F1 =    1.47
t =  0.100    F2 =   -1.12    M2 =   -2.01
d =  0.200    F3 =   -1.16    M3 =   -5.77
              F4 =   -1.52    M4 =   -4.13
              F5 =   -0.46    M5 =   -0.94
              F6 =    8.61
```

TABLE CPDEB 52   (9 FOLDS)

```
L = 14.760    F0 =    0.58
B =  7.160    F1 =    1.51
t =  0.100    F2 =   -1.17    M2 =   -2.09
d =  0.200    F3 =   -1.05    M3 =   -5.96
              F4 =   -1.47    M4 =   -4.28
              F5 =   -0.56    M5 =   -1.00
              F6 =    8.84
```

TABLE CPDEB 53   (9 FOLDS)

L = 14.760	F0 = 0.53		
B = 7.370	F1 = 1.55		
t = 0.100	F2 = -1.22	M2 =	-2.17
d = 0.200	F3 = -0.95	M3 =	-6.13
	F4 = -1.43	M4 =	-4.41
	F5 = -0.66	M5 =	-1.06
	F6 = 9.07		

TABLE CPDEB 54   (9 FOLDS)

L = 14.760	F0 = 0.48		
B = 7.590	F1 = 1.60		
t = 0.100	F2 = -1.29	M2 =	-2.26
d = 0.200	F3 = -0.84	M3 =	-6.29
	F4 = -1.39	M4 =	-4.54
	F5 = -0.77	M5 =	-1.12
	F6 = 9.31		

TABLE CPDEB 55   (9 FOLDS)

L = 14.760	F0 = 0.42		
B = 7.820	F1 = 1.65		
t = 0.100	F2 = -1.36	M2 =	-2.35
d = 0.200	F3 = -0.73	M3 =	-6.43
	F4 = -1.34	M4 =	-4.65
	F5 = -0.88	M5 =	-1.19
	F6 = 9.58		

TABLE CPDEB 56   (9 FOLDS)

L = 14.760	F0 = 0.37		
B = 8.050	F1 = 1.70		
t = 0.100	F2 = -1.43	M2 =	-2.45
d = 0.200	F3 = -0.62	M3 =	-6.54
	F4 = -1.30	M4 =	-4.74
	F5 = -0.99	M5 =	-1.26
	F6 = 9.85		

TABLE CPDEB 57   (9 FOLDS)

L = 15.200	F0 = 0.73	
B = 6.950	F1 = 1.49	
t = 0.100	F2 = −1.10	M2 = −2.02
d = 0.200	F3 = −1.27	M3 = −5.92
	F4 = −1.65	M4 = −4.27
	F5 = −0.40	M5 = −0.94
	F6 = 9.03	

TABLE CPDEB 58   (9 FOLDS)

L = 15.200	F0 = 0.67	
B = 7.160	F1 = 1.54	
t = 0.100	F2 = −1.15	M2 = −2.10
d = 0.200	F3 = −1.16	M3 = −6.13
	F4 = −1.60	M4 = −4.45
	F5 = −0.50	M5 = −1.00
	F6 = 9.26	

TABLE CPDEB 59   (9 FOLDS)

L = 15.200	F0 = 0.61	
B = 7.370	F1 = 1.58	
t = 0.100	F2 = −1.21	M2 = −2.18
d = 0.200	F3 = −1.06	M3 = −6.32
	F4 = −1.56	M4 = −4.60
	F5 = −0.60	M5 = −1.06
	F6 = 9.49	

TABLE CPDEB 60   (9 FOLDS)

L = 15.200	F0 = 0.56	
B = 7.590	F1 = 1.63	
t = 0.100	F2 = −1.27	M2 = −2.27
d = 0.200	F3 = −0.95	M3 = −6.50
	F4 = −1.51	M4 = −4.75
	F5 = −0.71	M5 = −1.12
	F6 = 9.75	

TABLE CPDEB 61   (9 FOLDS)

L = 15.200	F0 =	0.50		
B = 7.820	F1 =	1.68		
t = 0.100	F2 =	-1.34	M2 =	-2.36
d = 0.200	F3 =	-0.83	M3 =	-6.66
	F4 =	-1.46	M4 =	-4.88
	F5 =	-0.83	M5 =	-1.19
	F6 =	10.02		

TABLE CPDEB 62   (9 FOLDS)

L = 15.200	F0 =	0.45		
B = 8.050	F1 =	1.73		
t = 0.100	F2 =	-1.41	M2 =	-2.46
d = 0.200	F3 =	-0.71	M3 =	-6.80
	F4 =	-1.42	M4 =	-5.00
	F5 =	-0.94	M5 =	-1.26
	F6 =	10.30		

TABLE CPDEB 63   (9 FOLDS)

L = 15.200	F0 =	0.40		
B = 8.290	F1 =	1.78		
t = 0.100	F2 =	-1.49	M2 =	-2.57
d = 0.200	F3 =	-0.59	M3 =	-6.91
	F4 =	-1.38	M4 =	-5.10
	F5 =	-1.06	M5 =	-1.34
	F6 =	10.60		

TABLE CPDEB 64   (9 FOLDS)

L = 15.660	F0 =	0.76		
B = 7.160	F1 =	1.56		
t = 0.100	F2 =	-1.13	M2 =	-2.11
d = 0.200	F3 =	-1.28	M3 =	-6.29
	F4 =	-1.75	M4 =	-4.61
	F5 =	-0.44	M5 =	-1.00
	F6 =	9.71		

TABLE CPDEB 65 (9 FOLDS)

L = 15.660	F0 =	0.70		
B = 7.370	F1 =	1.61		
t = 0.100	F2 =	-1.19	M2 =	-2.19
d = 0.200	F3 =	-1.17	M3 =	-6.51
	F4 =	-1.70	M4 =	-4.79
	F5 =	-0.54	M5 =	-1.06
	F6 =	9.94		

TABLE CPDEB 66 (9 FOLDS)

L = 15.660	F0 =	0.65		
B = 7.590	F1 =	1.66		
t = 0.100	F2 =	-1.25	M2 =	-2.28
d = 0.200	F3 =	-1.06	M3 =	-6.71
	F4 =	-1.65	M4 =	-4.96
	F5 =	-0.65	M5 =	-1.12
	F6 =	10.21		

TABLE CPDEB 67 (9 FOLDS)

L = 15.660	F0 =	0.59		
B = 7.820	F1 =	1.71		
t = 0.100	F2 =	-1.32	M2 =	-2.37
d = 0.200	F3 =	-0.94	M3 =	-6.90
	F4 =	-1.60	M4 =	-5.12
	F5 =	-0.77	M5 =	-1.19
	F6 =	10.49		

TABLE CPDEB 68 (9 FOLDS)

L = 15.660	F0 =	0.53		
B = 8.050	F1 =	1.76		
t = 0.100	F2 =	-1.40	M2 =	-2.47
d = 0.200	F3 =	-0.82	M3 =	-7.06
	F4 =	-1.55	M4 =	-5.26
	F5 =	-0.89	M5 =	-1.26
	F6 =	10.78		

TABLE CPDEB 69   (9 FOLDS)

```
L =  15.660    F0 =     0.48
B =   8.290    F1 =     1.82
t =   0.100    F2 =    -1.48    M2 =    -2.58
d =   0.200    F3 =    -0.69    M3 =    -7.19
               F4 =    -1.51    M4 =    -5.38
               F5 =    -1.01    M5 =    -1.34
               F6 =    11.09
```

TABLE CPDEB 70   (9 FOLDS)

```
L =  15.660    F0 =     0.42
B =   8.540    F1 =     1.87
t =   0.100    F2 =    -1.56    M2 =    -2.69
d =   0.200    F3 =    -0.57    M3 =    -7.30
               F4 =    -1.46    M4 =    -5.49
               F5 =    -1.14    M5 =    -1.42
               F6 =    11.42
```

TABLE CPDEB 71   (9 FOLDS)

```
L =  16.130    F0 =     0.80
B =   7.380    F1 =     1.64
t =   0.100    F2 =    -1.16    M2 =    -2.20
d =   0.200    F3 =    -1.29    M3 =    -6.69
               F4 =    -1.84    M4 =    -4.98
               F5 =    -0.47    M5 =    -1.06
               F6 =    10.43
```

TABLE CPDEB 72   (9 FOLDS)

```
L =  16.130    F0 =     0.74
B =   7.600    F1 =     1.69
t =   0.100    F2 =    -1.23    M2 =    -2.29
d =   0.200    F3 =    -1.18    M3 =    -6.92
               F4 =    -1.79    M4 =    -5.17
               F5 =    -0.59    M5 =    -1.13
               F6 =    10.70
```

TABLE CPDEB 73 **(9 FOLDS)**

L = 16.130	F0 =	0.68		
B = 7.830	F1 =	1.74		
t = 0.100	F2 =	-1.30	M2 =	-2.39
d = 0.200	F3 =	-1.05	M3 =	-7.13
	F4 =	-1.74	M4 =	-5.36
	F5 =	-0.71	M5 =	-1.20
	F6 =	10.99		

TABLE CPDEB 74 **(9 FOLDS)**

L = 16.130	F0 =	0.62		
B = 8.060	F1 =	1.79		
t = 0.100	F2 =	-1.37	M2 =	-2.48
d = 0.200	F3 =	-0.93	M3 =	-7.32
	F4 =	-1.69	M4 =	-5.52
	F5 =	-0.83	M5 =	-1.27
	F6 =	11.29		

TABLE CPDEB 75 **(9 FOLDS)**

L = 16.130	F0 =	0.56		
B = 8.300	F1 =	1.85		
t = 0.100	F2 =	-1.46	M2 =	-2.59
d = 0.200	F3 =	-0.80	M3 =	-7.48
	F4 =	-1.64	M4 =	-5.67
	F5 =	-0.96	M5 =	-1.34
	F6 =	11.61		

TABLE CPDEB 76 **(9 FOLDS)**

L = 16.130	F0 =	0.50		
B = 8.550	F1 =	1.91		
t = 0.100	F2 =	-1.55	M2 =	-2.70
d = 0.200	F3 =	-0.66	M3 =	-7.61
	F4 =	-1.59	M4 =	-5.80
	F5 =	-1.09	M5 =	-1.43
	F6 =	11.95		

TABLE CPDEB 77   (9 FOLDS)

L =	16.130	F0 =	0.45			
B =	8.810	F1 =	1.97			
t =	0.100	F2 =	-1.64	M2 =	-2.83	
d =	0.200	F3 =	-0.53	M3 =	-7.72	
		F4 =	-1.55	M4 =	-5.92	
		F5 =	-1.22	M5 =	-1.51	
		F6 =	12.31			

TABLE CPDEB 78   (9 FOLDS)

L =	16.610	F0 =	0.84			
B =	7.600	F1 =	1.72			
t =	0.100	F2 =	-1.20	M2 =	-2.30	
d =	0.200	F3 =	-1.30	M3 =	-7.11	
		F4 =	-1.95	M4 =	-5.37	
		F5 =	-0.51	M5 =	-1.13	
		F6 =	11.20			

TABLE CPDEB 79   (9 FOLDS)

L =	16.610	F0 =	0.77			
B =	7.830	F1 =	1.77			
t =	0.100	F2 =	-1.27	M2 =	-2.39	
d =	0.200	F3 =	-1.18	M3 =	-7.35	
		F4 =	-1.89	M4 =	-5.58	
		F5 =	-0.63	M5 =	-1.20	
		F6 =	11.49			

TABLE CPDEB 80   (9 FOLDS)

L =	16.610	F0 =	0.71			
B =	8.060	F1 =	1.82			
t =	0.100	F2 =	-1.34	M2 =	-2.49	
d =	0.200	F3 =	-1.05	M3 =	-7.56	
		F4 =	-1.84	M4 =	-5.77	
		F5 =	-0.76	M5 =	-1.27	
		F6 =	11.80			

TABLE CPDEB 81 (9 FOLDS)

L = 16.610	F0 =	0.65		
B = 8.300	F1 =	1.88		
t = 0.100	F2 =	-1.43	M2 =	-2.60
d = 0.200	F3 =	-0.91	M3 =	-7.75
	F4 =	-1.79	M4 =	-5.94
	F5 =	-0.89	M5 =	-1.34
	F6 =	12.13		

TABLE CPDEB 82 (9 FOLDS)

L = 16.610	F0 =	0.59		
B = 8.550	F1 =	1.94		
t = 0.100	F2 =	-1.52	M2 =	-2.71
d = 0.200	F3 =	-0.77	M3 =	-7.92
	F4 =	-1.74	M4 =	-6.10
	F5 =	-1.03	M5 =	-1.43
	F6 =	12.49		

TABLE CPDEB 83 (9 FOLDS)

L = 16.610	F0 =	0.53		
B = 8.810	F1 =	2.00		
t = 0.100	F2 =	-1.62	M2 =	-2.84
d = 0.200	F3 =	-0.63	M3 =	-8.05
	F4 =	-1.69	M4 =	-6.25
	F5 =	-1.17	M5 =	-1.51
	F6 =	12.86		

TABLE CPDEB 84 (9 FOLDS)

L = 16.610	F0 =	0.48		
B = 9.070	F1 =	2.06		
t = 0.100	F2 =	-1.72	M2 =	-2.96
d = 0.200	F3 =	-0.49	M3 =	-8.15
	F4 =	-1.64	M4 =	-6.37
	F5 =	-1.31	M5 =	-1.60
	F6 =	13.24		

TABLE CPDEB 85　(9 FOLDS)

L = 17.110	F0 =	0.88		
B = 7.830	F1 =	1.80		
t = 0.100	F2 =	-1.23	M2 =	-2.40
d = 0.200	F3 =	-1.31	M3 =	-7.57
	F4 =	-2.06	M4 =	-5.80
	F5 =	-0.55	M5 =	-1.20
	F6 =	12.03		

TABLE CPDEB 86　(9 FOLDS)

L = 17.110	F0 =	0.81		
B = 8.060	F1 =	1.85		
t = 0.100	F2 =	-1.31	M2 =	-2.50
d = 0.200	F3 =	-1.18	M3 =	-7.80
	F4 =	-2.01	M4 =	-6.01
	F5 =	-0.67	M5 =	-1.27
	F6 =	12.35		

TABLE CPDEB 87　(9 FOLDS)

L = 17.110	F0 =	0.75		
B = 8.300	F1 =	1.91		
t = 0.100	F2 =	-1.39	M2 =	-2.61
d = 0.200	F3 =	-1.04	M3 =	-8.03
	F4 =	-1.95	M4 =	-6.22
	F5 =	-0.81	M5 =	-1.34
	F6 =	12.69		

TABLE CPDEB 88　(9 FOLDS)

L = 17.110	F0 =	0.68		
B = 8.550	F1 =	1.97		
t = 0.100	F2 =	-1.49	M2 =	-2.72
d = 0.200	F3 =	-0.89	M3 =	-8.22
	F4 =	-1.89	M4 =	-6.41
	F5 =	-0.95	M5 =	-1.43
	F6 =	13.05		

TABLE CPDEB 89   (9 FOLDS)

```
L =  17.110     F0 =     0.62
B =   8.810     F1 =     2.04
t =   0.100     F2 =    -1.59     M2 =    -2.84
d =   0.200     F3 =    -0.75     M3 =    -8.39
                F4 =    -1.84     M4 =    -6.58
                F5 =    -1.10     M5 =    -1.51
                F6 =    13.44
```

TABLE CPDEB 90   (9 FOLDS)

```
L =  17.110     F0 =     0.56
B =   9.070     F1 =     2.10
t =   0.100     F2 =    -1.70     M2 =    -2.97
d =   0.200     F3 =    -0.60     M3 =    -8.52
                F4 =    -1.79     M4 =    -6.73
                F5 =    -1.25     M5 =    -1.60
                F6 =    13.84
```

TABLE CPDEB 91   (9 FOLDS)

```
L =  17.110     F0 =     0.51
B =   9.340     F1 =     2.17
t =   0.100     F2 =    -1.80     M2 =    -3.11
d =   0.200     F3 =    -0.45     M3 =    -8.61
                F4 =    -1.74     M4 =    -6.86
                F5 =    -1.40     M5 =    -1.70
                F6 =    14.26
```

TABLE CPDEB 92   (9 FOLDS)

```
L =  17.620     F0 =     0.92
B =   8.060     F1 =     1.88
t =   0.100     F2 =    -1.27     M2 =    -2.51
d =   0.200     F3 =    -1.32     M3 =    -8.04
                F4 =    -2.18     M4 =    -6.25
                F5 =    -0.58     M5 =    -1.27
                F6 =    12.91
```

TABLE CPDEB 93   (9 FOLDS)

L = 17.620	F0 =	0.85		
B = 8.300	F1 =	1.94		
t = 0.100	F2 =	-1.35	M2 =	-2.61
d = 0.200	F3 =	-1.17	M3 =	-8.29
	F4 =	-2.12	M4 =	-6.48
	F5 =	-0.72	M5 =	-1.34
	F6 =	13.26		

TABLE CPDEB 94   (9 FOLDS)

L = 17.620	F0 =	0.78		
B = 8.550	F1 =	2.01		
t = 0.100	F2 =	-1.45	M2 =	-2.73
d = 0.200	F3 =	-1.02	M3 =	-8.52
	F4 =	-2.06	M4 =	-6.70
	F5 =	-0.87	M5 =	-1.43
	F6 =	13.63		

TABLE CPDEB 95   (9 FOLDS)

L = 17.620	F0 =	0.71		
B = 8.810	F1 =	2.07		
t = 0.100	F2 =	-1.56	M2 =	-2.85
d = 0.200	F3 =	-0.87	M3 =	-8.72
	F4 =	-2.01	M4 =	-6.90
	F5 =	-1.02	M5 =	-1.51
	F6 =	14.04		

TABLE CPDEB 96   (9 FOLDS)

L = 17.620	F0 =	0.65		
B = 9.070	F1 =	2.14		
t = 0.100	F2 =	-1.66	M2 =	-2.98
d = 0.200	F3 =	-0.71	M3 =	-8.88
	F4 =	-1.95	M4 =	-7.08
	F5 =	-1.18	M5 =	-1.60
	F6 =	14.45		

TABLE CPDEB 97   (9 FOLDS)

```
L =  17.620    F0 =    0.59
B =   9.340    F1 =    2.21
t =   0.100    F2 =   -1.78    M2 =   -3.12
d =   0.200    F3 =   -0.56    M3 =   -9.00
               F4 =   -1.90    M4 =   -7.24
               F5 =   -1.33    M5 =   -1.70
               F6 =   14.88
```

TABLE CPDEB 98   (9 FOLDS)

```
L =  17.620    F0 =    0.54
B =   9.620    F1 =    2.27
t =   0.100    F2 =   -1.89    M2 =   -3.26
d =   0.200    F3 =   -0.40    M3 =   -9.09
               F4 =   -1.85    M4 =   -7.39
               F5 =   -1.49    M5 =   -1.80
               F6 =   15.34
```

TABLE CPDEB 99   (9 FOLDS)

```
L =  18.150    F0 =    0.96
B =   8.300    F1 =    1.97
t =   0.100    F2 =   -1.31    M2 =   -2.62
d =   0.200    F3 =   -1.32    M3 =   -8.54
               F4 =   -2.31    M4 =   -6.74
               F5 =   -0.62    M5 =   -1.34
               F6 =   13.87
```

TABLE CPDEB 100   (9 FOLDS)

```
L =  18.150    F0 =    0.89
B =   8.550    F1 =    2.04
t =   0.100    F2 =   -1.40    M2 =   -2.73
d =   0.200    F3 =   -1.17    M3 =   -8.81
               F4 =   -2.25    M4 =   -6.99
               F5 =   -0.77    M5 =   -1.43
               F6 =   14.25
```

TABLE CPDEB 101   (9 FOLDS)

L =	18.150	F0 =	0.82		
B =	8.810	F1 =	2.11		
t =	0.100	F2 =	-1.51	M2 =	-2.86
d =	0.200	F3 =	-1.00	M3 =	-9.04
		F4 =	-2.19	M4 =	-7.23
		F5 =	-0.93	M5 =	-1.51
		F6 =	14.66		

TABLE CPDEB 102   (9 FOLDS)

L =	18.150	F0 =	0.75		
B =	9.070	F1 =	2.17		
t =	0.100	F2 =	-1.62	M2 =	-2.98
d =	0.200	F3 =	-0.84	M3 =	-9.24
		F4 =	-2.13	M4 =	-7.44
		F5 =	-1.09	M5 =	-1.60
		F6 =	15.09		

TABLE CPDEB 103   (9 FOLDS)

L =	18.150	F0 =	0.69		
B =	9.340	F1 =	2.25		
t =	0.100	F2 =	-1.74	M2 =	-3.12
d =	0.200	F3 =	-0.68	M3 =	-9.40
		F4 =	-2.07	M4 =	-7.63
		F5 =	-1.26	M5 =	-1.70
		F6 =	15.54		

TABLE CPDEB 104   (9 FOLDS)

L =	18.150	F0 =	0.63		
B =	9.620	F1 =	2.32		
t =	0.100	F2 =	-1.87	M2 =	-3.27
d =	0.200	F3 =	-0.51	M3 =	-9.52
		F4 =	-2.01	M4 =	-7.81
		F5 =	-1.42	M5 =	-1.80
		F6 =	16.02		

TABLE CPDEB 105   (9 FOLDS)

L = 18.150	F0 = 0.57		
B = 9.910	F1 = 2.39		
t = 0.100	F2 = -1.99	M2 =	-3.43
d = 0.200	F3 = -0.34	M3 =	-9.60
	F4 = -1.96	M4 =	-7.96
	F5 = -1.59	M5 =	-1.91
	F6 = 16.52		

TABLE CPDEB 106   (9 FOLDS)

L = 18.690	F0 = 1.00		
B = 8.550	F1 = 2.07		
t = 0.100	F2 = -1.35	M2 =	-2.74
d = 0.200	F3 = -1.31	M3 =	-9.09
	F4 = -2.44	M4 =	-7.28
	F5 = -0.67	M5 =	-1.43
	F6 = 14.89		

TABLE CPDEB 107   (9 FOLDS)

L = 18.690	F0 = 0.93		
B = 8.810	F1 = 2.14		
t = 0.100	F2 = -1.46	M2 =	-2.86
d = 0.200	F3 = -1.15	M3 =	-9.36
	F4 = -2.38	M4 =	-7.55
	F5 = -0.83	M5 =	-1.51
	F6 = 15.31		

TABLE CPDEB 108   (9 FOLDS)

L = 18.690	F0 = 0.86		
B = 9.070	F1 = 2.21		
t = 0.100	F2 = -1.57	M2 =	-2.99
d = 0.200	F3 = -0.98	M3 =	-9.59
	F4 = -2.31	M4 =	-7.79
	F5 = -0.95	M5 =	-1.60
	F6 = 15.75		

TABLE CPDEB 109   (9 FOLDS)

L = 18.690	F0 = 0.79			
B = 9.340	F1 = 2.28			
t = 0.100	F2 = -1.70	M2 = -3.13		
d = 0.200	F3 = -0.81	M3 = -9.79		
	F4 = -2.25	M4 = -8.02		
	F5 = -1.16	M5 = -1.70		
	F6 = 16.21			

TABLE CPDEB 110   (9 FOLDS)

L = 18.690	F0 = 0.72			
B = 9.620	F1 = 2.36			
t = 0.100	F2 = -1.83	M2 = -3.27		
d = 0.200	F3 = -0.63	M3 = -9.95		
	F4 = -2.19	M4 = -8.22		
	F5 = -1.34	M5 = -1.80		
	F6 = 16.71			

TABLE CPDEB 111   (9 FOLDS)

L = 18.690	F0 = 0.66			
B = 9.910	F1 = 2.43			
t = 0.100	F2 = -1.96	M2 = -3.43		
d = 0.200	F3 = -0.45	M3 = -10.06		
	F4 = -2.14	M4 = -8.41		
	F5 = -1.52	M5 = -1.91		
	F6 = 17.24			

TABLE CPDEB 112   (9 FOLDS)

L = 18.690	F0 = 0.60			
B = 10.210	F1 = 2.51			
t = 0.100	F2 = -2.10	M2 = -3.60		
d = 0.200	F3 = -0.27	M3 = -10.13		
	F4 = -2.09	M4 = -8.58		
	F5 = -1.70	M5 = -2.03		
	F6 = 17.78			

TABLE CPDEB 113   (9 FOLDS)

L =	19.250	F0 =	1.04	
B =	8.810	F1 =	2.17	
t =	0.100	F2 =	-1.40	M2 = -2.87
d =	0.200	F3 =	-1.31	M3 = -9.67
		F4 =	-2.59	M4 = -7.86
		F5 =	-0.71	M5 = -1.51
		F6 =	15.99	

TABLE CPDEB 114   (9 FOLDS)

L =	19.250	F0 =	0.97	
B =	9.070	F1 =	2.24	
t =	0.100	F2 =	-1.51	M2 = -2.99
d =	0.200	F3 =	-1.13	M3 = -9.94
		F4 =	-2.52	M4 = -8.14
		F5 =	-0.88	M5 = -1.60
		F6 =	16.44	

TABLE CPDEB 115   (9 FOLDS)

L =	19.250	F0 =	0.90	
B =	9.340	F1 =	2.32	
t =	0.100	F2 =	-1.64	M2 = -3.13
d =	0.200	F3 =	-0.95	M3 = -10.17
		F4 =	-2.45	M4 = -8.40
		F5 =	-1.05	M5 = -1.70
		F6 =	16.92	

TABLE CPDEB 116   (9 FOLDS)

L =	19.250	F0 =	0.83	
B =	9.620	F1 =	2.40	
t =	0.100	F2 =	-1.78	M2 = -3.28
d =	0.200	F3 =	-0.77	M3 = -10.37
		F4 =	-2.39	M4 = -8.64
		F5 =	-1.24	M5 = -1.80
		F6 =	17.43	

TABLE CPDEB 117   (9 FOLDS)

L = 19.250	F0 =	0.76		
B = 9.910	F1 =	2.48		
t = 0.100	F2 =	-1.92	M2 =	-3.43
d = 0.200	F3 =	-0.58	M3 =	-10.53
	F4 =	-2.33	M4 =	-8.86
	F5 =	-1.43	M5 =	-1.91
	F6 =	17.98		

TABLE CPDEB 118   (9 FOLDS)

L = 19.250	F0 =	0.70		
B = 10.210	F1 =	2.56		
t = 0.100	F2 =	-2.06	M2 =	-3.60
d = 0.200	F3 =	-0.39	M3 =	-10.64
	F4 =	-2.27	M4 =	-9.07
	F5 =	-1.62	M5 =	-2.03
	F6 =	18.55		

TABLE CPDEB 119   (9 FOLDS)

L = 19.250	F0 =	0.64		
B = 10.520	F1 =	2.64		
t = 0.100	F2 =	-2.21	M2 =	-3.79
d = 0.200	F3 =	-0.19	M3 =	-10.69
	F4 =	-2.22	M4 =	-9.25
	F5 =	-1.81	M5 =	-2.16
	F6 =	19.15		

TABLE CPDEB 120   (9 FOLDS)

L = 19.830	F0 =	1.09		
B = 9.070	F1 =	2.27		
t = 0.100	F2 =	-1.45	M2 =	-3.00
d = 0.200	F3 =	-1.30	M3 =	-10.28
	F4 =	-2.74	M4 =	-8.48
	F5 =	-0.75	M5 =	-1.60
	F6 =	17.17		

TABLE CPDEB 121   (9 FOLDS)

L = 19.830	F0 = 1.01		
B = 9.340	F1 = 2.35		
t = 0.100	F2 = -1.57	M2 = -3.13	
d = 0.200	F3 = -1.11	M3 = -10.55	
	F4 = -2.67	M4 = -8.78	
	F5 = -0.93	M5 = -1.70	
	F6 = 17.66		

TABLE CPDEB 122   (9 FOLDS)

L = 19.830	F0 = 0.94		
B = 9.620	F1 = 2.43		
t = 0.100	F2 = -1.71	M2 = -3.28	
d = 0.200	F3 = -0.92	M3 = -10.80	
	F4 = -2.60	M4 = -9.06	
	F5 = -1.12	M5 = -1.80	
	F6 = 18.19		

TABLE CPDEB 123   (9 FOLDS)

L = 19.830	F0 = 0.87		
B = 9.910	F1 = 2.52		
t = 0.100	F2 = -1.86	M2 = -3.44	
d = 0.200	F3 = -0.72	M3 = -10.99	
	F4 = -2.54	M4 = -9.32	
	F5 = -1.32	M5 = -1.91	
	F6 = 18.75		

TABLE CPDEB 124   (9 FOLDS)

L = 19.830	F0 = 0.80		
B = 10.210	F1 = 2.60		
t = 0.100	F2 = -2.02	M2 = -3.61	
d = 0.200	F3 = -0.52	M3 = -11.14	
	F4 = -2.47	M4 = -9.56	
	F5 = -1.52	M5 = -2.03	
	F6 = 19.35		

TABLE CPDEB 125   (9 FOLDS)

L = 19.830	F0 =	0.74		
B = 10.520	F1 =	2.69		
t = 0.100	F2 =	-2.17	M2 =	-3.79
d = 0.200	F3 =	-0.31	M3 =	-11.24
	F4 =	-2.41	M4 =	-9.78
	F5 =	-1.72	M5 =	-2.16
	F6 =	19.97		

TABLE CPDEB 126   (9 FOLDS)

L = 19.830	F0 =	0.68		
B = 10.840	F1 =	2.77		
t = 0.100	F2 =	-2.33	M2 =	-3.98
d = 0.200	F3 =	-0.11	M3 =	-11.28
	F4 =	-2.36	M4 =	-9.98
	F5 =	-1.92	M5 =	-2.29
	F6 =	20.62		

TABLE CPDEB 127   (9 FOLDS)

L = 20.420	F0 =	1.14		
B = 9.340	F1 =	2.38		
t = 0.100	F2 =	-1.50	M2 =	-3.14
d = 0.200	F3 =	-1.28	M3 =	-10.92
	F4 =	-2.90	M4 =	-9.15
	F5 =	-0.79	M5 =	-1.70
	F6 =	18.42		

TABLE CPDEB 128   (9 FOLDS)

L = 20.420	F0 =	1.06		
B = 9.620	F1 =	2.47		
t = 0.100	F2 =	-1.64	M2 =	-3.28
d = 0.200	F3 =	-1.08	M3 =	-11.21
	F4 =	-2.83	M4 =	-9.47
	F5 =	-0.99	M5 =	-1.80
	F6 =	18.96		

TABLE CPDEB 129   (9 FOLDS)
```
L =  20.420    F0 =     0.98
B =   9.910    F1 =     2.55
t =   0.100    F2 =    -1.79    M2 =   -3.44
d =   0.200    F3 =    -0.87    M3 = -11.45
               F4 =    -2.76    M4 =  -9.77
               F5 =    -1.19    M5 =  -1.91
               F6 =    19.54
```

TABLE CPDEB 130   (9 FOLDS)
```
L =  20.420    F0 =     0.91
B =  10.210    F1 =     2.64
t =   0.100    F2 =    -1.96    M2 =   -3.61
d =   0.200    F3 =    -0.66    M3 = -11.65
               F4 =    -2.69    M4 = -10.05
               F5 =    -1.40    M5 =  -2.03
               F6 =    20.16
```

TABLE CPDEB 131   (9 FOLDS)
```
L =  20.420    F0 =     0.84
B =  10.520    F1 =     2.74
t =   0.100    F2 =    -2.12    M2 =   -3.79
d =   0.200    F3 =    -0.45    M3 = -11.79
               F4 =    -2.63    M4 = -10.31
               F5 =    -1.62    M5 =  -2.16
               F6 =    20.81
```

TABLE CPDEB 132   (9 FOLDS)
```
L =  20.420    F0 =     0.78
B =  10.840    F1 =     2.83
t =   0.100    F2 =    -2.29    M2 =   -3.98
d =   0.200    F3 =    -0.23    M3 = -11.87
               F4 =    -2.56    M4 = -10.54
               F5 =    -1.83    M5 =  -2.29
               F6 =    21.49
```

TABLE CPDEB 133   (9 FOLDS)

L = 20.420	F0 = 0.72	
B = 11.170	F1 = 2.91	
t = 0.100	F2 = -2.46	M2 = -4.19
d = 0.200	F3 = -0.01	M3 = -11.90
	F4 = -2.51	M4 = -10.76
	F5 = -2.04	M5 = -2.43
	F6 = 22.20	

TABLE CPDEB 134   (9 FOLDS)

L = 21.030	F0 = 1.18	
B = 9.620	F1 = 2.50	
t = 0.100	F2 = -1.56	M2 = -3.28
d = 0.200	F3 = -1.26	M3 = -11.61
	F4 = -3.08	M4 = -9.87
	F5 = -0.84	M5 = -1.80
	F6 = 19.78	

TABLE CPDEB 135   (9 FOLDS)

L = 21.030	F0 = 1.10	
B = 9.910	F1 = 2.59	
t = 0.100	F2 = -1.72	M2 = -3.44
d = 0.200	F3 = -1.04	M3 = -11.90
	F4 = -3.00	M4 = -10.22
	F5 = -1.05	M5 = -1.91
	F6 = 20.37	

TABLE CPDEB 136   (9 FOLDS)

L = 21.030	F0 = 1.02	
B = 10.210	F1 = 2.68	
t = 0.100	F2 = -1.88	M2 = -3.61
d = 0.200	F3 = -0.82	M3 = -12.15
	F4 = -2.93	M4 = -10.54
	F5 = -1.27	M5 = -2.03
	F6 = 21.00	

TABLE CPDEB 137   (9 FOLDS)

L = 21.030	F0 = 0.95	
B = 10.520	F1 = 2.78	
t = 0.100	F2 = -2.06	M2 = -3.79
d = 0.200	F3 = -0.60	M3 = -12.34
	F4 = -2.86	M4 = -10.84
	F5 = -1.49	M5 = -2.16
	F6 = 21.68	

TABLE CPDEB 138   (9 FOLDS)

L = 21.030	F0 = 0.88	
B = 10.840	F1 = 2.88	
t = 0.100	F2 = -2.24	M2 = -3.98
d = 0.200	F3 = -0.37	M3 = -12.47
	F4 = -2.79	M4 = -11.12
	F5 = -1.72	M5 = -2.29
	F6 = 22.39	

TABLE CPDEB 139   (9 FOLDS)

L = 21.030	F0 = 0.82	
B = 11.170	F1 = 2.97	
t = 0.100	F2 = -2.42	M2 = -4.19
d = 0.200	F3 = -0.13	M3 = -12.53
	F4 = -2.73	M4 = -11.37
	F5 = -1.94	M5 = -2.43
	F6 = 23.13	

TABLE CPDEB 140   (9 FOLDS)

L = 21.030	F0 = 0.76	
B = 11.510	F1 = 3.06	
t = 0.100	F2 = -2.60	M2 = -4.41
d = 0.200	F3 = 0.10	M3 = -12.54
	F4 = -2.67	M4 = -11.60
	F5 = -2.17	M5 = -2.58
	F6 = 23.90	

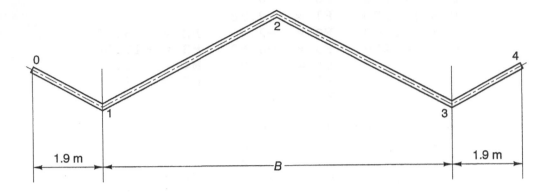

Fig. A9

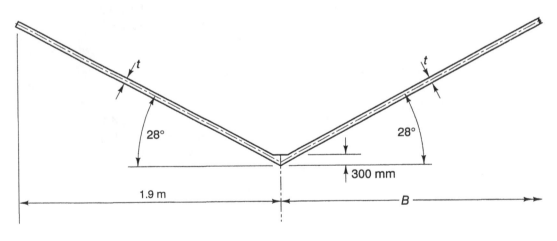

Fig. A10

Appendix 9. Design tables for folded plates types UEC[1]

```
TABLE UEC 1    (3 FOLDS)
L =  18.000    F0 =    -6.66
B =   8.240    F1 =     5.69    M1 =    -8.62
t =   0.120    F2 =    -5.24    M2 =    -8.58


TABLE UEC 2    (3 FOLDS)
L =  18.000    F0 =    -6.69
B =   8.490    F1 =     5.58    M1 =    -8.62
t =   0.120    F2 =    -5.09    M2 =    -9.11


TABLE UEC 3    (3 FOLDS)
L =  18.000    F0 =    -6.72
B =   8.740    F1 =     5.48    M1 =    -8.62
t =   0.120    F2 =    -4.94    M2 =    -9.66


TABLE UEC 4    (3 FOLDS)
L =  18.000    F0 =    -6.75
B =   9.000    F1 =     5.38    M1 =    -8.62
t =   0.120    F2 =    -4.81    M2 =   -10.26


TABLE UEC 5    (3 FOLDS)
L =  18.000    F0 =    -6.79
B =   9.270    F1 =     5.29    M1 =    -8.62
t =   0.120    F2 =    -4.67    M2 =   -10.90


TABLE UEC 6    (3 FOLDS)
L =  18.000    F0 =    -6.83
B =   9.550    F1 =     5.19    M1 =    -8.62
t =   0.120    F2 =    -4.55    M2 =   -11.59
```

[1] See Figs A9 and A10

TABLE UEC 7   (3 FOLDS)
L = 18.000   F0 =   -6.86
B = 9.840   F1 =   5.10   M1 = -8.62
t = 0.120   F2 =   -4.42   M2 = -12.33

TABLE UEC 8   (3 FOLDS)
L = 18.540   F0 =   -6.55
B = 8.240   F1 =   5.66   M1 = -9.60
t = 0.140   F2 =   -5.25   M2 = -11.05

TABLE UEC 9   (3 FOLDS)
L = 18.540   F0 =   -6.59
B = 8.490   F1 =   5.56   M1 = -9.60
t = 0.140   F2 =   -5.10   M2 = -11.59

TABLE UEC 10   (3 FOLDS)
L = 18.540   F0 =   -6.62
B = 8.740   F1 =   5.46   M1 = -9.60
t = 0.140   F2 =   -4.95   M2 = -12.16

TABLE UEC 11   (3 FOLDS)
L = 18.540   F0 =   -6.66
B = 9.000   F1 =   5.36   M1 = -9.60
t = 0.140   F2 =   -4.82   M2 = -12.78

TABLE UEC 12   (3 FOLDS)
L = 18.540   F0 =   -6.70
B = 9.270   F1 =   5.27   M1 = -9.60
t = 0.140   F2 =   -4.68   M2 = -13.44

TABLE UEC 13   (3 FOLDS)
L = 18.540   F0 =   -6.74
B =   9.550   F1 =   5.18   M1 =   -9.60
t =   0.140   F2 =   -4.55   M2 = -14.16

TABLE UEC 14   (3 FOLDS)
L = 18.540   F0 =   -6.78
B =   9.840   F1 =   5.09   M1 =   -9.60
t =   0.140   F2 =   -4.43   M2 = -14.95

TABLE UEC 15   (3 FOLDS)
L = 19.100   F0 =   -6.93
B =   8.240   F1 =   6.00   M1 =   -9.60
t =   0.140   F2 =   -5.57   M2 = -11.60

TABLE UEC 16   (3 FOLDS)
L = 19.100   F0 =   -6.96
B =   8.490   F1 =   5.89   M1 =   -9.60
t =   0.140   F2 =   -5.41   M2 = -12.13

TABLE UEC 17   (3 FOLDS)
L = 19.100   F0 =   -7.00
B =   8.740   F1 =   5.79   M1 =   -9.60
t =   0.140   F2 =   -5.26   M2 = -12.68

TABLE UEC 18   (3 FOLDS)
L = 19.100   F0 =   -7.04
B =   9.000   F1 =   5.69   M1 =   -9.60
t =   0.140   F2 =   -5.11   M2 = -13.28

TABLE UEC 19   (3 FOLDS)
L = 19.100   F0 =   -7.09
B =   9.270   F1 =   5.59   M1 =   -9.60
t =   0.140   F2 =   -4.97   M2 = -13.93

TABLE UEC 20   (3 FOLDS)
L = 19.100   F0 =   -7.13
B =   9.550   F1 =   5.49   M1 =   -9.60
t =   0.140   F2 =   -4.83   M2 = -14.63

TABLE UEC 21   (3 FOLDS)
L = 19.100   F0 =   -7.18
B =   9.840   F1 =   5.39   M1 =   -9.60
t =   0.140   F2 =   -4.70   M2 = -15.40

TABLE UEC 22   (3 FOLDS)
L = 19.670   F0 =   -7.31
B =   8.240   F1 =   6.35   M1 =   -9.60
t =   0.140   F2 =   -5.91   M2 = -12.21

TABLE UEC 23   (3 FOLDS)
L = 19.670   F0 =   -7.35
B =   8.490   F1 =   6.24   M1 =   -9.60
t =   0.140   F2 =   -5.74   M2 = -12.72

TABLE UEC 24   (3 FOLDS)
L = 19.670   F0 =   -7.40
B =   8.740   F1 =   6.13   M1 =   -9.60
t =   0.140   F2 =   -5.58   M2 = -13.26

TABLE UEC 25 **(3 FOLDS)**
L = 19.670   F0 =   -7.44
B = 9.000   F1 =   6.02   M1 =   -9.60
t = 0.140   F2 =   -5.42   M2 = -13.84

TABLE UEC 26 **(3 FOLDS)**
L = 19.670   F0 =   -7.49
B = 9.270   F1 =   5.92   M1 =   -9.60
t = 0.140   F2 =   -5.27   M2 = -14.47

TABLE UEC 27 **(3 FOLDS)**
L = 19.670   F0 =   -7.54
B = 9.550   F1 =   5.81   M1 =   -9.60
t = 0.140   F2 =   -5.13   M2 = -15.15

TABLE UEC 28 **(3 FOLDS)**
L = 19.670   F0 =   -7.59
B = 9.840   F1 =   5.71   M1 =   -9.60
t = 0.140   F2 =   -4.99   M2 = -15.89

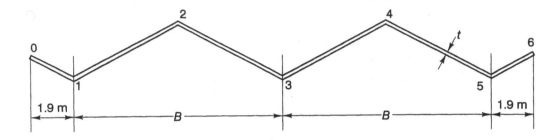

Fig. A11

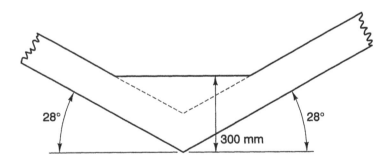

Fig. A12

Appendix 10. Design tables for folded plates types CUEC[1]

```
TABLE CUEC 1   (5 FOLDS)
L =  18.000    F0 =   -8.29
B =   8.240    F1 =    5.98    M1 =   -8.62
t =   0.120    F2 =   -4.56    M2 =   -8.15
               F3 =    4.20    M3 =   -6.33

TABLE CUEC 2   (5 FOLDS)
L =  18.000    F0 =   -8.27
B =   8.490    F1 =    5.86    M1 =   -8.62
t =   0.120    F2 =   -4.44    M2 =   -8.60
               F3 =    4.09    M3 =   -6.70

TABLE CUEC 3   (5 FOLDS)
L =  18.000    F0 =   -8.26
B =   8.740    F1 =    5.75    M1 =   -8.62
t =   0.120    F2 =   -4.32    M2 =   -9.06
               F3 =    3.98    M3 =   -7.08

TABLE CUEC 4   (5 FOLDS)
L =  18.000    F0 =   -8.24
B =   9.000    F1 =    5.64    M1 =   -8.62
t =   0.120    F2 =   -4.21    M2 =   -9.57
               F3 =    3.88    M3 =   -7.49

TABLE CUEC 5   (5 FOLDS)
L =  18.000    F0 =   -8.24
B =   9.270    F1 =    5.53    M1 =   -8.62
t =   0.120    F2 =   -4.09    M2 =  -10.12
               F3 =    3.78    M3 =   -7.93

TABLE CUEC 6   (5 FOLDS)
L =  18.000    F0 =   -8.23
B =   9.550    F1 =    5.42    M1 =   -8.62
t =   0.120    F2 =   -3.99    M2 =  -10.72
               F3 =    3.67    M3 =   -8.39
```

[1] See Figs A11 and A12

TABLE CUEC 7   (5 FOLDS)

L = 18.000	F0 = -8.23	
B = 9.840	F1 = 5.31	M1 = -8.62
t = 0.120	F2 = -3.88	M2 = -11.37
	F3 = 3.57	M3 = -8.89

TABLE CUEC 8   (5 FOLDS)

L = 18.540	F0 = -8.22	
B = 8.240	F1 = 5.98	M1 = -9.60
t = 0.140	F2 = -4.59	M2 = -10.15
	F3 = 4.24	M3 = -6.68

TABLE CUEC 9   (5 FOLDS)

L = 18.540	F0 = -8.21	
B = 8.490	F1 = 5.86	M1 = -9.60
t = 0.140	F2 = -4.47	M2 = -10.61
	F3 = 4.13	M3 = -7.10

TABLE CUEC 10   (5 FOLDS)

L = 18.540	F0 = -8.20	
B = 8.740	F1 = 5.74	M1 = -9.60
t = 0.140	F2 = -4.35	M2 = -11.09
	F3 = 4.02	M3 = -7.53

TABLE CUEC 11   (5 FOLDS)

L = 18.540	F0 = -8.19	
B = 9.000	F1 = 5.63	M1 = -9.60
t = 0.140	F2 = -4.23	M2 = -11.61
	F3 = 3.91	M3 = -8.00

TABLE CUEC 12   (5 FOLDS)

L = 18.540	F0 = -8.19	
B = 9.270	F1 = 5.52	M1 = -9.60
t = 0.140	F2 = -4.12	M2 = -12.18
	F3 = 3.81	M3 = -8.49

TABLE CUEC 13 (5 FOLDS)
```
L = 18.540    F0 =   -8.19
B =  9.550    F1 =    5.42   M1 =   -9.60
t =  0.140    F2 =   -4.01   M2 = -12.80
              F3 =    3.71   M3 =   -9.02
```

TABLE CUEC 14 (5 FOLDS)
```
L = 18.540    F0 =   -8.19
B =  9.840    F1 =    5.31   M1 =   -9.60
t =  0.140    F2 =   -3.90   M2 = -13.48
              F3 =    3.61   M3 =   -9.59
```

TABLE CUEC 15 (5 FOLDS)
```
L = 19.100    F0 =   -8.71
B =  8.240    F1 =    6.34   M1 =   -9.60
t =  0.140    F2 =   -4.87   M2 = -10.56
              F3 =    4.50   M3 =   -6.54
```

TABLE CUEC 16 (5 FOLDS)
```
L = 19.100    F0 =   -8.69
B =  8.490    F1 =    6.21   M1 =   -9.60
t =  0.140    F2 =   -4.74   M2 = -10.99
              F3 =    4.38   M3 =   -6.96
```

TABLE CUEC 17 (5 FOLDS)
```
L = 19.100    F0 =   -8.68
B =  8.740    F1 =    6.09   M1 =   -9.60
t =  0.140    F2 =   -4.61   M2 = -11.45
              F3 =    4.26   M3 =   -7.40
```

TABLE CUEC 18 (5 FOLDS)
```
L = 19.100    F0 =   -8.68
B =  9.000    F1 =    5.98   M1 =   -9.60
t =  0.140    F2 =   -4.49   M2 = -11.96
              F3 =    4.15   M3 =   -7.87
```

```
TABLE CUEC 19   (5 FOLDS)
L = 19.100    F0 =    -8.67
B =   9.270   F1 =     5.86   M1 =    -9.60
t =   0.140   F2 =    -4.37   M2 = -12.52
              F3 =     4.04   M3 =   -8.37

TABLE CUEC 20   (5 FOLDS)
L = 19.100    F0 =    -8.68
B =   9.550   F1 =     5.75   M1 =    -9.60
t =   0.140   F2 =    -4.26   M2 = -13.12
              F3 =     3.93   M3 =   -8.90

TABLE CUEC 21   (5 FOLDS)
L = 19.100    F0 =    -8.68
B =   9.840   F1 =     5.63   M1 =    -9.60
t =   0.140   F2 =    -4.14   M2 = -13.78
              F3 =     3.83   M3 =   -9.47

TABLE CUEC 22   (5 FOLDS)
L = 19.670    F0 =    -9.21
B =   8.240   F1 =     6.71   M1 =    -9.60
t =   0.140   F2 =    -5.17   M2 = -11.00
              F3 =     4.77   M3 =   -6.39

TABLE CUEC 23   (5 FOLDS)
L = 19.670    F0 =    -9.20
B =   8.490   F1 =     6.58   M1 =    -9.60
t =   0.140   F2 =    -5.03   M2 = -11.42
              F3 =     4.64   M3 =   -6.81

TABLE CUEC 24   (5 FOLDS)
L = 19.670    F0 =    -9.19
B =   8.740   F1 =     6.46   M1 =    -9.60
t =   0.140   F2 =    -4.89   M2 = -11.86
              F3 =     4.52   M3 =   -7.25
```

```
TABLE CUEC 25   (5 FOLDS)
L = 19.670    F0 =   -9.19
B =  9.000    F1 =    6.33   M1 =   -9.60
t =  0.140    F2 =   -4.76   M2 = -12.35
              F3 =    4.40   M3 =   -7.73

TABLE CUEC 26   (5 FOLDS)
L = 19.670    F0 =   -9.18
B =  9.270    F1 =    6.21   M1 =   -9.60
t =  0.140    F2 =   -4.64   M2 = -12.89
              F3 =    4.29   M3 =   -8.23

TABLE CUEC 27   (5 FOLDS)
L = 19.670    F0 =   -9.18
B =  9.550    F1 =    6.09   M1 =   -9.60
t =  0.140    F2 =   -4.51   M2 = -13.48
              F3 =    4.17   M3 =   -8.77

TABLE CUEC 28   (5 FOLDS)
L = 19.670    F0 =   -9.19
B =  9.840    F1 =    5.97   M1 =   -9.60
t =  0.140    F2 =   -4.39   M2 = -14.12
              F3 =    4.06   M3 =   -9.34
```

Index

Printed and bound by CPI Group (UK) Ltd, Croydon, CR0 4YY

01/11/2024

01782602-0002